75)
→79

80-81
81-86

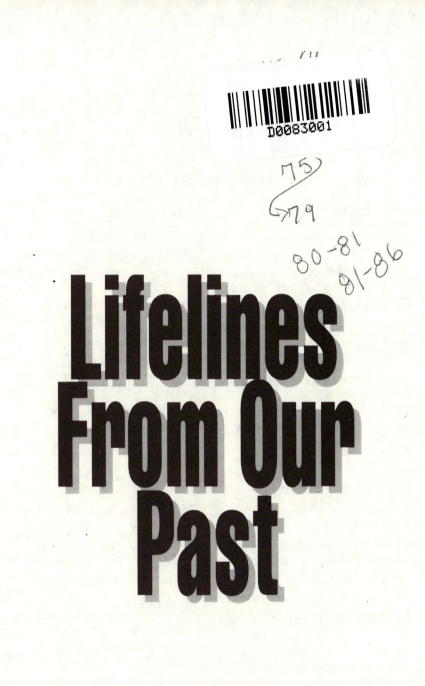

Lifelines From Our Past

Sources and Studies in World History

Kevin Reilly, Series Editor

THE ALCHEMY OF HAPPINESS
Abu Hamid Muhammad al-Ghazzali
Translated by Claud Field, revised and annotated
by Elton L. Daniel

NATIVE AMERICANS BEFORE 1492
The Moundbuilding Centers of the
Eastern Woodlands
Lynda Norene Shaffer

GERMS, SEEDS, AND ANIMALS
Studies in Ecological History
Alfred W. Crosby

BALKAN WORLDS
The First and Last Europe
Traian Stoianovich

AN ATLAS AND SURVEY OF SOUTH
ASIAN HISTORY
Karl J. Schmidt

THE GOGO
History, Customs, and Traditions
Mathias E. Mnyampala
Translated, introduced, and edited by
Gregory H. Maddox

WOMEN IN WORLD HISTORY:
Volume 1—Readings from Prehistory to 1500
Volume 2—Readings from 1500 to the Present
Edited by Sarah Shaver Hughes and
Brady Hughes

MARITIME SOUTHEAST ASIA TO 1500
Lynda Norene Shaffer

THE COURSE OF HUMAN HISTORY
Economic Growth, Social Process, and
Civilization
Johan Goudsblom, Eric Jones, and Stephen
Mennell

ON WORLD HISTORY
An Anthology
Johann Gottfried Herder
Translated by Ernest A. Menze with
Michael Palma, edited by Hans Adler and Ernest
A. Menze

TEACHING WORLD HISTORY
A Resource Book
Edited by Heidi Roupp

DOUBLE GHOSTS
Oceanian Voyages on Euroamerican Ships
David A. Chappell

SHAPING WORLD HISTORY
Breakthroughs in Ecology, Technology, Science,
and Politics
Mary Kilbourne Matossian

TRADITION AND DIVERSITY
Christianity in a World Context to 1500
Edited by Karen Louise Jolly

LIFELINES FROM OUR PAST
A New World History
Revised Edition
L.S. Stavrianos

THE FACE OF THE EARTH
Environment and World History
Edited by J. Donald Hughes

THE WORLD THAT TRADE
CREATED
Society, Culture, and the World Economy,
1400 to the Present
Kenneth Pomeranz and Steven Topik

COLONIALISM AND THE MODERN
WORLD
Selected Studies
Edited by Gregory Blue, Martin Bunton, and
Ralph Croizier

THE POWER OF SCALE
A Global History Approach
John H. Bodley

RACISM
A Global Reader
Edited by Kevin Reilly, Stephen Kaufman,
and Angela Bodino

THE WORLD AND A VERY SMALL PLACE
IN AFRICA (2ND ED.)
A History of Globalization in Niumi, The Gambia
Donald R. Wright

Sources
and
Studies
in World
History

L.S. Stavrianos

Lifelines From Our Past

A New World History

Revised Edition

M.E. Sharpe
Armonk, New York
London, England

Copyright © 1989, 1997 by L. S. Stavrianos
Foreword copyright © 2004 by M. E. Sharpe, Inc.
80 Business Park Drive, Armonk, New York 10504

Grateful acknowledgement is made to the following for permission to reprint previously
published material:
B. T. Batsford Limited: Excerpt from *Everyday Life in Early Imperial China* by Michael
Loewe. Reprinted by permission of the publisher, B. T. Batsford Limited, London.
International African Institute: Excerpt from *Land, Labour, and Diet in Northern
Rhodesia* by A. I. Richards. Originally published by Oxford University Press for the
International African Institute. Reprinted by permission.

Book Design by Michael Mendelsohn of M 'N O Production Services, Inc.

Library of Congress Cataloging-in-Publication Data

Stavrianos, Leften Stavros.
Lifelines from our past : a new world history / L. S. Stavrianos—Rev. ed.
p. cm. — (Sources and studies in world history)
Includes bibliographical references and index.
ISBN 0-7656-0180-X (pbk. : alk. paper)
1. World history. I. Title. II. Series.
D20.S83 1997
909—dc21
97-12880
CIP

Printed in the United States of America

The paper used in this publication meets the minimum
requirements of American National Standard for
Information Sciences—Permanence of Paper for
Printed Library Materials, ANSI Z 39.48-1984.

∞

BM (p) 10 9 8 7

CONTENTS

Foreword by Kevin Reilly   ix

Acknowledgments   xi

Introduction: Lifelines from My Past   3

1. KINSHIP SOCIETIES   15

2. TRIBUTARY SOCIETIES   43

3. CAPITALIST SOCIETIES   87

4. HUMAN PROSPECTS   189

5. WORLD HISTORY FOR THE TWENTY-FIRST CENTURY   251

Notes   265

Index   275

To the memory of B.K.S.,
who is the coauthor in more ways
than she might have imagined.

FOREWORD

When L. S. Stavrianos passed away this spring, we lost our country's, and perhaps the world's, earliest and most influential author and advocate of world history in our schools and colleges. Between 1960 and 2000, probably more students were introduced to world history through one of the texts or anthologies of Stavrianos than through those of anyone else. Many of us who teach world history today first learned from him the power of the big picture clearly presented. For decades, to teach world history was to teach the texts of Stavrianos or William H. McNeill.

It was fitting that two Canadian scholars in Balkan history initiated world history in America after World War II. Canada provided the necessary perspective. The Balkans offered a world in miniature, numerous languages to master, and the need to produce a narrative that transcended national needs. After publishing in Greek and Balkan history, both created models of world history and received foundation support to write the first school and college texts.

Both were materialists, Stavrianos out of Marx and a line of evolutionary anthropology that stretched back to Lewis Henry Morgan; McNeill a student of technology, demography, and ecology. Each concentrated on their own kind of social history. McNeill charted the impact of the tools of war, pathogens, and the interaction of steppe and sown; Stavrianos studied political power and social class. Perhaps they called on world history to do different things. Both wanted it to explain, but Stavrianos also wanted to change the world.

"I have worlds to conquer," he would tell his children in the morning. Part Alexander and part country fisherman, he was a bear of a man, a cosmopolitan Zorba, eating yogurt after a swim in his La Jolla pool while scheming to save the world with his pen. Those who have had the privilege of knowing the man personally know that Leften Stavrianos wrote as he spoke: passionately and directly. Neither shy nor guarded, Leften seemed almost constitutionally incapable of mincing his words. But the format of textbook publishing often calls for balance in lieu of passion. In consequence, his works for a more general audience, books like *The Coming Dark Ages* and *The Global Rift*, were often more fun to read.

Lifelines from Our Past, his last and most engaging book for the general reader, has also been widely used in the classroom. The reasons are clear. In fewer than half the pages usually taken by a text, the author presents a disarmingly simple, yet penetratingly profound, vision of the entire history of humanity. Eliminating much of the tedious detail that historians often feel compelled to recount, Stavrianos gives us, instead, a survival course. He attempts to tell us about the most important changes in the human past so that we can build the most fulfilling future.

The result is both daring and persuasive. He calls our attention to the importance of the major technological and economic revolutions in world history: the agricultural revolution that transformed age old paleolithic patterns ten thousand years ago and the string of capitalist revolutions (commercial, industrial, and high-tech) that have transformed the world in the last few hundred years. Each revolution, he finds, has surpassed the previous one in both creative and destructive effects. Each has called for a radical reassessment of the human condition.

His combination of lavish brushstrokes and deftly chosen quotations both startles and alarms us. *Lifelines* rings like a firebell in the night, but it also offers the understanding that we will need for renewal.

Kevin Reilly
July 2004

ACKNOWLEDGMENTS

This book is the end result of a lifetime of quarrying information and ideas from myriad sources. Space forbids specific enumeration, so I content myself with a comprehensive bow, and with mentioning by name only those who read and criticized portions of the manuscript: Elman R. Service (University of California, Santa Barbara), Mary E. Clark (San Diego State University), and Allen Lein (University of California, San Diego). I am especially indebted to my colleague, John W. Dower (University of California, San Diego), who, despite the pressure of other commitments, read the entire manuscript and responded with a detailed and penetrating critique which, despite what must have seemed at the time to be perversity on my part, I deeply appreciated then as I do now. I owe the greatest debt to my editor, Tom Engelhardt, who was as unsparing with his comments as with his time, and who prodded me to rethink and revise the manuscript from beginning to end. It was a much-appreciated experience, as enlightening as it was unique for this author.

The staff of the UCSD Central University Library responded unfailingly to repeated requests throughout the preparation of this manuscript. I am indebted also to Jacqueline Griffin for her conscientiousness and skill in typing and retyping the manuscripts in its several reincarnations as it shuttled back and forth between the Pacific and Atlantic coasts.

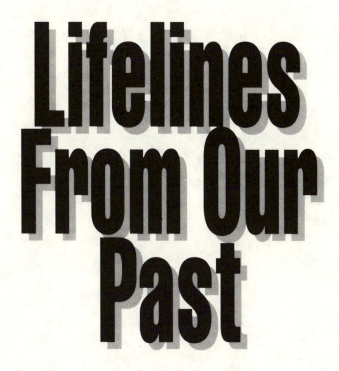

Lifelines From Our Past

INTRODUCTION

LIFELINES FROM MY PAST

Each generation must write its history for itself, and draw its own deductions from that already written; not because the conclusions of its predecessors are untrue, but for a practical reason. Different answers are required, because different questions are asked. Standing at a new point on the road, it finds that fresh ranges of the landscape come into view, whose unfamiliar intricacies demand an amplification of traditional charts.

R. H. TAWNEY

All macrohistory is autobiography.

The roots of this particular macrohistory stretch back to the Depression when, as a teenager, I worked as a waiter in a skid-row restaurant to help stock the lean family larder. The restaurant was located in the beautiful city of Vancouver. There my impressionable young mind was confronted daily with the contrast between the destitution and misery of the customers I waited on and the visibly boundless riches of the province of British Columbia.

Most of the restaurant's customers had been laborers on farms or in forests, mines, or fisheries. But in these Depression years, many of their customary jobs had vanished. Instead of earned wages, they were forced to subsist on weekly meal-ticket vouchers available at provincial relief offices. With these vouchers in hand, the unemployed then trooped to the skid-row restaurants where they could get the most out of their few pennies.

Since British Columbia has the mildest winter weather of any Canadian province, Canada's unemployed flocked to Vancouver (the only large city on the Canadian Pacific coast), just as America's unemployed flocked to California. Every day they came by the hundreds, jumping off freight trains and heading for the relief offices. Our restaurant's daily waves of new customers could be timed to the schedule of freight trains pulling into Vancouver from the frozen prairie provinces.

It was obvious even to the boy I was then that a large percentage of the unemployed were alcoholics, and that many had been maimed while working in factories, mines, or logging camps, on trawlers, or in canneries. Very common were missing fingers, broken or severed legs or arms, and patches over eyeless sockets. The skid row in which these men lived, and in which I worked weekends and summer "vacations" for several years, had all the tawdry landmarks common to such locales: cheap rooming houses, pawnshops, beer parlors, and assorted hustlers.

Altogether different, and belligerently vocal, were the Wobbly orators who spoke every evening in the square on which our restaurant happened to front—Vancouver's version of Hyde Park. Usually, after their soapbox performances, the speakers would come into the restaurant for coffee and pie. The hour being late by then, and the customers few, I would be able to stop for a cup of coffee and a little talk with those spokesmen for the IWW—Industrial Workers of the World. I was overwhelmed by what they told me— by their scorn for "pie in the sky" promises, by their dismissal of city, provincial, and dominion officials as "flunkies" of ruling elites rather than servants of the people, by their scornful references to the commonly admired "Mounties" (Royal Canadian Mounted Police) as "Cossacks," and by their claim that Canada's transcontinental railways, with whose freight trains they were intimately familiar, should be operated and owned by the workers who had built them, not the bankers and stockholders who profited from them.

Heady stuff for a bewildered adolescent, but no more so than the reality that surrounded me. With the particular earnestness of those years of life, I strove to sort it all out immediately, searching for answers to the questions raised not only by the Wobblies but by what I was witnessing and experiencing daily. One central question

4

would not leave me: why the cruel gap between the glittering potential and the wretched reality of beautiful Vancouver? I knew that my customers and friends hated to be out of work and yearned for jobs. It was obvious that the millions who were ill fed, ill clothed, and ill housed urgently needed the food that was not being grown, the lumber that was not being milled, and the fish that was not being canned. Why, then, were the unemployed blocked from the work for which they were trained, and for which Canadian society had such a naked need? Why?

To this simple and distressing question I found myself returning again and again, despite the allure of Vancouver's beaches and mountains, which I somehow managed to enjoy with youthful zest. If the Wobblies' answers stunned but in the end did not satisfy me, at least they spurred me to ask questions and to challenge proffered answers in a quest for intellectual and emotional understanding.

That quest led me initially to speculation which, in retrospect, seems embarrassingly simpleminded. One theory that intrigued me at the time was that Canada's problems might stem from the absence in our history of any revolution analogous to the American one. The United States was a wealthy country, and Americans were relatively prosperous (or so we Depression-buffeted Canadians erroneously believed at the time). Since the Americans had successfully overthrown British imperial rule in the eighteenth century, and since Canadians still remained within the British Empire, it seemed to me that perhaps the way out for us beleaguered imperial subjects was to proclaim our own Declaration of Independence. This hypothesis had the added attraction of conforming to what the Wobblies had told me about the way British interests were exploiting Canada's natural and human resources.

While such youthful speculation may appear naive today, those skid-row experiences that engendered it left an imprint far deeper and more enduring than anything I ever learned in a classroom or textbook. However clichéd it may sound, that skid-row restaurant was my first and foremost "university." And to my mind, what I learned there remains compelling to the present—that all societies are powerfully flawed by a gap between official rhetoric and the social reality, between promise and practice, and that the role of a historian should be to cast light on the origins of that gap.

This seemed, even then, in the context of my personal experiences, self-evident and incontrovertible, so I was taken aback to discover that my teachers by no means shared my view. When I asked them point-blank why they had opted to teach history (for that was the discipline I, not surprisingly, chose for undergraduate and graduate study), they offered vague nostrums about the need for educated people to be familiar with their cultural heritage. This response seemed, in the light of my personal experiences and observations, a cop-out, and I rejected it. So I persisted in my resolve to study history in order to use it as a tool for social understanding rather than as a cultural ornament.

The completion of graduate studies in 1937 was followed by an academic career of a full half-century's duration in several Canadian and American universities. The specific subjects on which I focused my research and writing reflected my enduring skid-row "legacy." In graduate school, I specialized in the history of the Balkan Peninsula, regarded by nineteenth-century writers as the underdeveloped Third World of the time. Because of the recurring wars, revolutions, and international crises originating in the peninsula, it was commonly referred to as the "powder keg of Europe." Although the powder-keg analogy was fully justified, it was little emphasized that the European Great Powers (tsarist Russia, Hohenzollern Germany, Hapsburg Austria-Hungary, and imperial Britain) were purposefully supplying the "gunpowder" in pursuit of their respective interests. The small Balkan states were essentially pawns, moved about on the diplomatic chessboard in much the same way many small Central American and Southeast Asian states have been manipulated in our own times.

I also noted that on those exceptional occasions when the Balkan states were able to subordinate their national and religious rivalries to the cause of peninsular cooperation and unity, the Great Powers invariably intervened to "divide and rule," playing off one Balkan people against another. I decided to study Balkan history from the inside rather than from the lofty perspective of European chancelleries. My doctoral dissertation, a history of the Balkan federation movement, together with the later publication of a full-scale history, *The Balkans Since 1453*, gained for me the professional status of a "Balkan expert."

But a life of Balkan expertise hardly measured up to the tumultuous, rapidly changing world outside the classroom. A tidal wave of post–World War II colonial revolutions was sweeping away the centuries-old European empires. Closer to home, students at Northwestern University, where I was then teaching, were being taken for service in Korea, and it was painfully obvious that they had little notion of where they were going or why. The more I thought about it, the more I realized that this was symptomatic of the all-pervasive Western orientation of American scholarship, not just in history and not just at Northwestern, but in all disciplines and on all campuses.

My own response was to try to cope with this distortion where it was most flagrant in my own discipline—that is, in the obligatory introductory history course for incoming freshmen. Almost invariably, this course was a history of "Western civilization," in fact and often in name. Teacher and text began in Egypt and Mesopotamia, crossed the Mediterranean to classical Greece and Rome, crossed the Alps to northern Europe, and finally crossed the Atlantic to the New World, where civilization, of course, attained its full flowering. How few of us understood that in pursuing this ever-westward course of "civilization" we had managed to exclude the great majority of the human race in Asia, Africa, and the pre-Columbian Americas from our students' sight. We bothered with Japan only to set the stage for Admiral Perry, with India only in order to introduce Robert Clive, and with Africa only to herald the arrival of David Livingstone and Cecil Rhodes.

This exclusivity did not even strike most of us as exclusive because of the way in which we had taken the West's prewar global hegemony for granted. Within the context of this distorted balance of power, the traditional Western orientation of history teaching seemed only natural and proper. But with the postwar colonial revolutions, the old context quickly and visibly crumbled, as the changing names and colors on any map readily indicated. Slowly and reluctantly, some of us began to recognize the obsolescence of our ways of organizing the world for our students.

At this time, I made a decision to leave the field of Balkan history and to venture forth into world history, which was then not generally accepted as a legitimate field for research and teaching. Specifically, I set out to organize a new introductory course that would be genuinely global in scope and yet pedagogically manageable. This involved experimenting with the excision of many time-honored topics appropriate for a Western history but hardly central to a one-year global survey.

Thanks to the backing of the Carnegie Corporation of New York, I was afforded the considerable time necessary to develop the fundamentally new concepts and structures that were needed. Most basic was the proposition that world history was not and could not be merely the sum of national or regional histories, that it must instead be dealt with as a new integral whole comprising more than the sum of its historical parts.

A few other historians were also beginning to think and work along these lines in the 1950s and 1960s. As a result, a new global approach was presented in books and articles, as well as in courses at the elementary, secondary, and college levels. Gradually, the global perspective won a following, a convincing indicator being the rapidly growing number of new world-history textbooks being issued by publishers eager to cash in on an expanding market. A decisive factor behind the success of this approach was, of course, the impact of events then taking place in Vietnam, southern Africa, the Middle East, and elsewhere, which established beyond any doubt the need for broader pedagogical horizons.

If the legitimacy and need for world history is today no longer seriously questioned, a much more basic issue has arisen in recent years. The new issue is not whether there should be world history but rather what sort of world history it should be. Our world today is as different from that of the 1950s, when the movement for world history began, as the world of the fifties was from the pre–World War II era. Just as the global destabilization caused by the colonial revolutions, the Korean and Vietnam wars (and the "Cold" War) necessitated a new globally oriented history, so further innovation is now dictated by even more basic destabilizing developments, including environmental degradation, resource depletion, global structural unemployment, increasing hunger amid increasing plenty, and above all, the specter of nuclear winter. Once again, we face the need to rethink and reframe history in terms applicable to our current experiences and needs.

The issue is no longer simply West versus non-West, or a Toynbeean assessment of the meaning of the rise and fall of civilizations, or even of the rise and fall of modern great powers (as recently analyzed by Paul Kennedy), however important and useful these efforts have been.[1] The issue, rather, involves ourselves as a species—the nature of our human nature, values, achievements, and prospects.

Paradoxically, the need for such integral stocktaking arises from the fact that we as a species have always been the great overachievers of planet Earth. So overwhelming has been our "success" that we are now in the process of superseding billions of years of evolution by natural selection with a new era of evolution by human selection. We are determining not only what happens to our planet and to its plant and animal life, but even what happens to the genes of those plants and animals (including our own). Evolution is no longer determined simply by genes—or by natural extinction-causing cataclysms. Rather, it is increasingly determined by humans, and consequently has been transformed into a qualitatively new type of "directed" process, directed at least in the sense that the fate of the earth and its various inhabitants is in our hands, whether we fully and consciously accept that responsibility or not. As physicist Werner Heisenberg points out: "For the first time in the course of history, man on earth faces only himself; he finds no longer any other partner or foe."[2]

It is hardly too dramatic to insist that we are finally face to face with ourselves. No longer can we avoid asking why this age of unprecedented human dominance and achievement is also the age when the possibility of species extinction for the first time is a sober possibility.

The South African statesman Jan Christiaan Smuts once declared, "When I look at history I am a pessimist . . . but when I look at prehistory, I am an optimist."[3] This observation is not only on target in terms of history but of history-writing itself—for our prehistory has often and unreasonably been scanted in assessing the human past. It is as if attitudes and behaviors in prehistoric eras could simply be dismissed as somehow less than human. Yet our "history" began only with the first emergence of civilization—and writing—in Mesopotamia less than six thousand years ago, while the variety of human experience extends back to *Homo sapiens*'s emergence some fifty thousand to one hundred thousand years ago as indicated by the latest archeological findings in the Middle East. And if we look further back to our hominid origins, we must extend our time frame some 5 million years to encompass our Australopithecine ancestors. Although the recorded events of the past few thousand years have naturally dominated our histories and our consciousness, the fact remains that prehistory, which after all is well over 99 percent of our human heritage, offers quite a different, and often a far more hopeful, window into human possibilities, even in terms of the future we face.

In recent years, anthropologists have made fundamental discoveries concerning the lives of our prehistoric ancestors. These have helped rid us of certain shibboleths, rationalizations, and prejudices concerning "human nature," "race," and other crucial subjects about which so much nonsense has been written with so much certitude. Instead, we can now appraise, with accumulating evidence and reasonable assurance, the nature of human nature and of basic human values, not to speak of the successes and failures human societies have had in meeting basic human needs over a span of tens of thousands of years rather than over a few short centuries. The nature of human nature and how humans have dealt with each other as well as the world around them are the bedrock topics that cannot be evaded in any responsible appraisal of human prospects in the late twentieth century. These are the "lifeline" issues that need to be defined and traced from our past, prehistoric as well as historic.

Granted such a need, there still arises the problem of viability. Whereas Toynbee's civilizations encompassed a few thousand years, and Kennedy's great powers five centuries, prehistory extends back—depending on one's definition—as far as five million years. How can the evolution of our species, spanning such a time frame, be summarized and interpreted meaningfully within the confines of a book like this? A simple and serviceable starting point is provided by the fact that humans are social animals who have always lived within social environments of their own making. These social environments, or societies, have appeared all over the globe in countless varieties, depending on such local factors as climate (so that Eskimo and Arab societies differed), physical setting (so that Andean and Polynesian societies differed), and degree of accessibility to outside influences (so that the isolated Australian aborigines and the centrally located Middle Eastern peoples had entirely different historical experiences).

All these factors conjoined in innumerable permutations and combinations to produce the grand total of past and present human societies. Since that total is formidable, the task of appraising those societies and relating them to our times and needs is correspondingly formidable. It becomes more manageable only when it is noted that all of these hundreds of past and present human societies fall into three broad categories: kinship societies, encompassing all human communities until about 3500 B.C.; tributary societies (also known as civilizations), which appeared first in the Middle East about 3500 B.C. and gradually diffused to or emerged autonomously on all continents except Australia; and free-market, or capitalist, societies, which first appeared in northwestern Europe about 1500 A.D. (although the precise date is in dispute), and expanded until they encompassed and dominated the entire globe.

All three social systems proved long-lasting because they met certain basic human needs of their time. But in each case, sooner or later, as conditions changed, certain contradictions developed within the systems, so that they became increasingly dysfunctional until each (except our own to date) gave way to a new type of society that resolved the contradiction. How the original needs were met, how contradictions appeared, and how they were resolved in the past is instructive for us today, for we are experiencing the same basic paradigm, though at a greatly accelerated tempo and heightened intensity because of the driving force of capitalism and in particular of the high technology it has developed.

Thanks to this common pattern in human affairs, it is possible to pick up strands of Ariadne's thread that ran through all three societal types, to discern certain basic lifeline issues that allow us to explore and compare past and present human experiments and experiences. For the sake of clarity, and at the obvious risk of oversimplification, these "lifelines" are analyzed for each of the three social systems under four headings: ecology, gender relations, social relations, and war. These four topics are selected not because numerous others are deemed insignificant, but because they encompass the broadest areas of human experience of immediate concern to us today and allow the maximum linkage between our past and our present global crisis.

It might be objected that such a brief and highly selective survey is not really world history, that it does not portray the human epic in its full complexity and rich color. Indeed, it does not, which makes it necessary to define precisely what *Lifelines from Our Past* does do.

It is definitely not a comprehensive history of the human past. Several such histories have appeared recently and are generally available.[4] This book is instead a highly selective analysis of those aspects of the past that illuminate our present. It is, in short, an inquiry into our usable past. Of course, the quest for a usable past should involve society's collective experiences and reflections as they coalesce into a shared consensus providing guidance for the future. Yet it is necessary to recognize certain pitfalls that beset the pursuit of relevance in history. Relevance does not signify the teleological assumption of some overarching design or purpose in the human past. Relevance also does not necessarily mean "recent." That Paleolithic people worked much less than we do today, and probably suffered proportionally less starvation and malnutrition than our present 5 billion do, is of infinitely more relevance than most information we get on our nightly news reports.

It should be especially emphasized that "relevance" in a historian's thinking should not be confused with "prediction." We cannot look to history for precise answers to the problems of the present or to those we imagine we will face in the future. History, of course, is not an exact science like physics or chemistry. History deals with human beings whose actions can hardly be predicted at all, much less with the certainty that a chemist can predict what will happen when element A is combined with element B. It cannot be used like a crystal ball to predict what political party will win, what national leader will be assassinated, which country will have a revolution, or where a war will break out. On the other hand, it does not follow that history sheds no light on the relations between past, present, and future. If we purposefully look for meaningful patterns, and are competent and responsible in our search, then history becomes a useful discipline. Its usefulness is not in being predictive, but in providing a framework for considering past and present—a framework that will not foretell what is to come, but that can reveal the human flexibility and human potentiality that is our legacy.

Finally, if history is not crystal-gazing, so it is not bare chronicling of facts. Untenable is the common assumption that by piling fact upon fact the assembled facts will speak for themselves. They will not. The basic task we face today is not to accumulate still more factual data, but to make sense of the vast store of information we already possess.

For this reason the guiding watchword in preparing this study has been "Dare to Omit." Inevitably, the omissions include not just familiar and cherished historical figures and events, but significant subjects such as the state, nationalism, and the high cultures that figure prominently and properly in standard general histories.

If this sounds too much like a tract for our times, it should be recognized that the histories of all peoples in bygone ages also emerged as "tracts" for their respective times, addressing explicitly or implicitly their specific needs. It was natural that our prehistoric ancestors should have conceived of static creation myths since they assumed that everything, including themselves, their culture, and their habitat, had appeared with the creation and was destined to continue unaltered into the future. It was equally natural that the historians of medieval Christian Europe viewed their task as one of illustrating and justifying God's way to humans. They knew that civilization was under a sentence of death and therefore they wrote history primarily to warn the careless and to remind the frivolous of approaching judgment.

Natural also was the Western-oriented history of nineteenth-century scholars who viewed a world in which European empires encompassed the entire globe. The focus of those scholars understandably was not Jerusalem, but the preordained westward course of empire. Finally, it was natural that Western-oriented history should begin to be questioned following the two world wars and the colonial revolutions that dismantled those European empires. As the discrepancy between written history and the real world became uncomfortably conspicuous, the need for a new global approach to the human experience was gradually recognized and conceded.

The basic thesis of this book is that today still another review of human history has become imperative—a review from a new angle of vision reflecting the new facts and new needs of the late twentieth century. Each generation must write its own history, not because past histories are untrue but because in a rapidly changing world new questions arise and new answers are needed.

Asking new questions and formulating new answers is easier said than done. A major obstacle in the quest for a usable past is psychological, as John Maynard Keynes suggested in the introduction to his *General Theory of Employment, Interest, and Money* (1936). Keynes informs us that in writing that book he experienced a "long struggle of escape from habitual modes of thought and expression. . . . The ideas which are here expressed so laboriously are extremely simple and should be obvious. The difficulty lies not in the new ideas but in escaping from the old ones, which ramify, for those brought up as most of us have been, into every corner of our minds."

Keynes's point is particularly significant and relevant for historians. All intellectuals must wrestle with the numbing grip of the past, but historians especially so because of the very nature of their subject matter. What we already know keeps us from looking at the past through fresh eyes to learn what we need to know now.

Ralph Waldo Emerson defined the issue precisely when he observed that "the use of history is to give value to the present hour." This book is offered with Emerson's objective in mind, and with the warning to the reader that the challenge we face is to shed the old as much as it is to discover the new.

CHAPTER 1

KINSHIP SOCIETIES

When I look at history, I am a pessimist . . . but when I look at prehistory, I am an optimist.

JAN C. SMUTS

However crude and ineffectual primitive cultures were in their control over the forces of nature, they had worked out a system of human relationships that has never been equaled since the Agricultural Revolution. The warm, substantial bonds of kinship united man with man. There were no lords or vassals, serfs or slaves, in tribal society. In social ritual one man might make obeisance to another, but no one kept another in bondage and lived upon the fruits of his labor. There were no time clocks, no bosses or overseers, in primitive society, and a two-week vacation was not one's quota of freedom for a year. . . . Crude and limited as primitive cultures may have been technologically . . . their social systems based upon kinship and characterized by liberty, equality, and fraternity were unquestionably more congenial to the human primate's nature, and more compatible with his psychic needs and aspirations, than any other that has ever been realized in any of the cultures subsequent to the Agricultural Revolution, including our own society today.

LESLIE A. WHITE

HUMAN ORIGINS

From their beginnings, human beings have been the great revolutionaries of planet Earth. They started their turbulent history with a profoundly disruptive achievement that has differentiated them from all other creatures; they have continued their revolutionary ways to the present day.

The root cause of this "addiction" to revolution is the human brain. It has enabled humans to break with the traditional glacial mode of evolution by genetic adaptation to environment, and enter a fast lane in which environment is adapted to genes. More specifically, humans have used their superior brain to achieve two historic breakthroughs. One was their technology, the ways in which they learned to manipulate the physical environment to serve their own needs. The other was their social organization, which was superior to that of all other primates. The technological achievements are generally recognized, but not so the parallel achievements in social organization. Yet the latter have been at least as important in making human ascendancy to the status of global primacy possible. Today, in fact, how humans have organized themselves is of particular significance if we are to understand why the survival of the species is in question at the very moment of unprecedented technological triumphs.

Considering first the historic role of human technology, it should be noted that other animals employ tools. The California sea otter uses a stone under water to pound mollusks loose from rocks. Likewise, a chimpanzee will take a twig, trim it of leaves, and poke it into a termite hill. When pulled out after a few moments, the twig is usually covered with termites, which the chimpanzee licks off with relish. Thus, otters and chimpanzees as well as other animals must be classified along with humans as tool users.

Yet, compared to all other animals, there is an obvious qualitative difference in the degree to which humans make and use tools. The key to this difference is the human brain, which not only receives sensory impulses but also analyzes them, stores them as memory, and then synthesizes them. Thus, these impulses can be converted into a wide repertoire of behaviors, enabling humans to function, first and foremost, as generalists. They never adapted exclusively to one type of environment, as the gibbons with their long lithe arms did to the forest, or the antelopes with their fleet legs did to the open savanna, or the polar bears with their heavy white fur did to the Arctic. Rather, humans have adapted with their brain, which has allowed them to use all environments from the equator to the poles. Simultaneously, the brain allowed humans progressively to develop their technology (in contrast to apes whose technology is no more advanced today than it was a hundred thousand years ago).

The early technological achievements of *Homo sapiens*'s hominid ancestors included not only toolmaking but also the use of fire and the development of speech. These unique advances enabled the hominids to break out of the warm African savannas where they are generally believed to have originated. A succession of hominid lineages overlapped each other through the millennia, known to us now only by tongue-twisting names derived from fossil discoveries. *Australopithecus* dates back some 5 million years B.P. (before the present), while our immediate ancestor, *Homo erectus*, first appeared about 1.6 million years B.P. Out of this evolutionary process emerged archaic forms of *Homo sapiens* (thinking humans) about 500,000 B.P., and finally modern *Homo sapiens* about 50,000 B.P. The latter date may have to be pushed back to about 100,000 B.P. in light of recent fossil discoveries in Israel.[1]

By that date, the human species had populated all the continents except Antarctica. Even at the modest rate of ten miles a generation, our ancestors could have spread out from Nairobi to Peking in less than 15,000 years, a brief period from an evolutionary perspective. During the Ice Ages, ocean waters were sucked up into the ice caps, leaving exposed land bridges connecting the continents, or at least bringing them far closer than they are today. As a result, humans crossed the land bridge from Siberia to North America, now covered by the waters of the Bering Strait, and also the straits from Southeast Asia to Australia.

Hand in hand with the dispersal of *Homo sapiens* went race differentiation. The relative isolation of the scattered human communities combined with adaptation to different local environments to generate the variety of existing races with their distinctive skin colors, hair colors and forms, and face morphologies, including eye folds and nose and lip structures. The important point to note is that this race differentiation occurred well after the emergence of *Homo sapiens*, which explains why all races can interbreed, and why anthropologists agree that no significant differences in mental capacity exist among the races of humankind. Their widely different historical experiences stem not from differences in ability but from accidents of history.

FROM PRIMATE SOCIETY TO HUMAN SOCIETY

Technology was a critical factor in the success of *Homo sapiens*, whether in the tools they used to prevail against predators, in the use of fire to open new horizons in human diet and habitat, or in the development of speech that enabled each generation, thanks to its storehouse of language, to begin where the previous generation left off rather than having to start anew. Yet fully as significant for the human triumph, and particularly relevant for us today, was the type of social organization evolved by our early ancestors and their hominid predecessors.

Most prehistorians and anthropologists now accept the idea that the basic characteristic of this original human society was its foundation on kinship relations. In direct contrast to primate society, it was cooperative and communal. Kinship relations meant that when a paleolithic hunter returned to camp with the carcass of a deer, he shared it as a matter of course not only with his own immediate family but also with other members of his band who were all kinfolk of one sort or another. This sharing was taken for granted, not because some compensation would be forthcoming in the future, but because of the expectations that the kinship ties themselves engendered. Anthropologist Elman Service has drawn an analogy with our own society, where it is assumed that parents will "give" food to their children, and the children in turn will "help" their parents in their old age.[2]

The pervasiveness of this "generalized form of reciprocity" in food-gathering societies is illustrated by an experience of the Danish anthropologist and explorer Peter Freuchen. He thanked an Eskimo hunter for a gift of meat, only to be met by the hunter's obvious displeasure over the proffered thanks. An old man then explained to Freuchen the reason for the negative reaction: "You must not thank for your meat: it is your right to get parts. In this country, nobody wishes to be dependent on others. Therefore, there is nobody who gives or gets gifts, for thereby you become dependent. With gifts you make slaves just as with whips you make dogs."[3] Precisely this characteristic of sharing because it seems the natural and right thing to do has been found by anthropologists among the native peoples of Australia: "It is with [the Australian aborigine] a fixed habit to give away part of what he has, and he neither expects the man to whom he gives a thing to express his gratitude, nor, when a native gives him anything, does he think it necessary to do so himself, for the simple reason that giving and receiving are matters of course in his everyday life."[4]

Interpersonal relations based on kinship ties date back millions of years to early ancestors such as the Australopithecines and persisted as the dominant form of social interaction until the advent of tributary societies about 3500 B.C. In other words, kinship societies have been the overwhelmingly predominant human experience through the millennia and across the globe. Such chronological and geographic pervasiveness was not, however, the product of distinctive genes or even community calculation. It was simply the natural concomitant of the Paleolithic way of life. Communal sharing went hand in hand with common ownership and access to food supplies freely available in the surrounding countryside. For Paleolithic food gatherers, their natural environment served as an ever-full and ever-available refrigerator. Whenever a band found the local supply of animal and plant food becoming exhausted, they simply moved on to a fresh campsite. As a result, paleolithic bands were ever on the move, literally eating their way out of one campsite and into the next.

Recent archaeological excavations reveal that in exceptional cases where local plant and animal food resources were unusually rich, it was possible for food gatherers to settle down in year-round villages, even though the inhabitants subsisted entirely on hunting and gathering. This was the case at Abu Hureyra in northern Syria, where wild cereals, pulses, and legumes grew so densely that they yielded harvests as rich as though they were in planted gardens, and where the annual migration of Persian gazelles provided a plentiful and reliable meat supply. This favorable combination enabled a large village of about three-to-four hundred to flourish from 9500 B.C. to 8100 B.C., when climatic change reduced the rich steppe flora and forced the villagers to move on. Similar exceptional circumstances enabled other food gatherers to subsist in permanent settlements, as in the Pacific Northwest, where bountiful fish resources were available year-round. Therefore, anthropologists now conclude that our early ancestors were not everywhere restricted to a pattern of either nomadic hunting and gathering, or sedentary farming. The pattern did vary from site to site, depending on specific local conditions, so that good-sized villages existed for lengthy periods without plant or animal domestication. Yet the fact remains that such villages were the exception. Nomadism was the natural corollary of food gathering, just as a settled life was of food producing.[5]

The nomadic mode of life made individual accumulation not only unnecessary but impractical. Material possessions had to be strictly limited when, every few weeks or months, everything had to be gathered up and carried to the next camping area. Under such conditions, the impulse to acquire and accumulate, which we assume to be inherent in "human nature," was literally unthinkable. This helps to explain the irritation so common in Western observers when they perceived (from the perspective of their own acquisitive societies) the "improvidence" and "irresponsibility" of the kinship peoples they discovered overseas.

Kinship social organization has prevailed through almost all of human history not only because it was the most functional for nomadic food gatherers, but also because it best served their survival needs in potentially harsh physical environments. Cooperative kinship relations between humans proved superior to the more competitive relationships often found in primate social life, as is clear in the contrasting behaviors regulating the basic drives for sex and food.

In ape society, competition in the quest for females is settled by fighting or at least mock displays of aggressiveness, from which the strongest and most aggressive males emerge dominant. Each of these males rules over several females and their young, while weaker or younger males are forced to the fringes of the ape social system. By contrast, humans regulate their sex relations through the institution of marriage, a socially recognized pairing arrangement. Instead of sex being a disruptive force as in ape society, it plays a binding or even peacemaking role among humans. Incest taboos, for instance, promote the choosing of mates from neighboring bands, thereby establishing friendly personal ties between groups over a large area.

Likewise, in the distribution of food, the dominant male among apes allows others to eat only after he is sated. Again, there was a sharply contrasting situation in early human societies where all historical and present-day observation of kinship peoples tells us that food was brought back to the band base by both male hunters and female gatherers for communal sharing, cooking, and eating. In human kinship society, food, like sex, enhanced not friction and competition as in ape society, but cooperation and sociability.

Anthropologists are agreed that it was precisely this cooperative feature of early human society that ensured the survival and eventual success of the species. There is no evidence that this was due to some unique human gene or genes for altruism. Individual human beings, as anthropologist Elman Service comments, are as self-serving as other animals. According to Service, the secret of *Homo sapiens*'s success is that "human societies have systems of rewards and punishments and associated joys and fears that can make service to one's fellows become simultaneously a self-service."[6] The same basic point is made by anthropologist Marshall Sahlins:

In selective adaptation to the perils of the Stone Age, human society overcame or subordinated such primate propensities as selfishness, indiscriminate sexuality, dominance and brute competition. It substituted kinship and cooperation for conflict, placed solidarity over sex, morality over might. In its earliest days it accomplished the greatest reform in history, the overthrow of human primate nature, and thereby secured the evolutionary future of the species. . . . The emerging human primate, in a life-and-death economic struggle with nature, could not afford the luxury of a social struggle. Cooperation, not competition, was essential.[7]

If the cooperative, communal features of Paleolithic society contributed fundamentally to the early human struggle for survival against environmental dangers, then this has possible implications for the current struggle of the species for survival in the man-made environment that has to a large extent replaced the natural. Relevant in this connection is the experience of the arctic explorer Vilhjalmur Stefansson, who lived with the Eskimos of Coronation Gulf in Canada's Northwest Territories between 1906 and 1918. He noted that all Eskimos had access to community supplies of food, clothing, and other necessities, and that it was not necessary to accumulate for old age because "the community will support you gladly when you are too old to work." Stefansson also observed that status depended upon "your judgment, your ability, and your character, but notably upon your unselfishness and kindness." Significantly, Stefansson concluded from his long involvement with the Eskimos "that the chief factor in their happiness was that they were living according to the Golden Rule. Man is more fundamentally a cooperative animal than a competitive animal. His survival as a species has been through mutual aid rather than through rugged individualism."[8]

THE KUNG

Since human behavior, unlike bones, does not become fossilized, anthropologists must rely mainly on contemporary food gatherers when they seek to determine the nature of Paleolithic society. But can we assume that our Paleolithic ancestors had the same social organization as today's food gatherers, simply because they, too, lived off the bounty of nature rather than producing their own food. Anthropologists believe the assumption can be made because of the basic similarity of all food-gathering societies today, regardless of whether they are located in the Arctic or the Amazon, the deserts of Australia or southern Africa.

The remarkably similar ways in which all kinship societies function despite the radically different environments they inhabit suggests that the determining factor in their lives is the limited set of alternatives open to food gatherers, regardless of their geographic location or the period in which they thrive. All peoples living off the land face basically similar problems and so evolve roughly similar social institutions. Probably the principal difference between Paleolithic food gatherers and those of today is that the latter have been driven into undesirable peripheral regions—deserts and jungles, for example—where they are subsisting under the most difficult of conditions. Their Paleolithic ancestors, by contrast, had access to the entire globe, including the fertile regions with hospitable climates that are now populated by more numerous and more powerful agricultural and industrial peoples. Consequently, the food-gathering societies observed by anthropologists today cannot be considered ideal representatives of their genre; rather they are societies that have somehow survived against overwhelming odds, are now hanging on under the most adverse and stressful conditions, and face the prospect of an even less favorable future.

Given these dismal circumstances, it is all the more significant that anthropologists in recent years have found it necessary to abandon the traditional Hobbesian view of food-gathering life as "solitary, poor, nasty, brutish, and short."[9] Today each one of these adjectives has been replaced by its exact opposite. Food-gathering society is now viewed as "the original affluent society," whose members work "bankers' hours" and enjoy healthy diets, economic security, and a warm social life.[10] This reappraisal is based on studies of surviving bands on all continents, the most detailed being those of the Kung group of the Bushmen living in the Kalahari Desert of southern Africa. Since 1963 the Kung have been carefully studied by anthropologists, archaeologists, linguists, psychologists, and nutritionists. Their findings jibe with those from other continents, and together yield a revealing and significant insight into the mode of life that prevailed during more than 95 percent of human history.

One surprising revelation to emerge from studies of the Kung is how abundant and reliable are their food supplies, despite an unfavorable environment. This is due in part to their extraordinary knowledge of their home territory and all its plant and animal life. Although these nomads cannot read or write, they can learn and remember—so much so that it is estimated that their fund of information, transmitted orally from generation to generation, would fill thousands of volumes.

The Kung use no less than five hundred species of plants and animals as food, or for medical, cosmetic, toxic, and other purposes. Between 60 and 80 percent of Kung food is obtained by the women, who gather plants (bulbs, beans, roots, leafy greens, berries and nuts, especially the all-important mongongo fruit and nut) as well as small mammals, tortoises, snakes, caterpillars, insects, and bird eggs. Although Westerners are culturally programmed to reject most of these food sources, the fact is that beetle grubs, caterpillars, bee larvae, termites, ants, and cicadas, all of which are eaten today by gatherers, are highly nutritious. Termites, for instance, are about 45 percent protein, a higher proportion than that found even in protein-rich dried fish. Men contribute to the Kung diet by hunting animals, snaring birds, and extracting honey from beehives. The diversity of these food sources ensures a year-round reliable supply of food even under the most adverse climatic conditions, in contrast to agriculturists, who must depend on the few crops they grow, and therefore are vulnerable to droughts, frosts, floods, and pests. In fact, anthropologists noted that during a serious drought in the summer of 1964, the Kung food supply remained as plentiful as usual, while the neighboring Bantu farmers starved. To feed their hungry families, Bantu women joined their Kung sisters in their foraging expeditions.

Not only abundant and dependable, the Kung food supply also constitutes an exceptionally healthy diet. It is low in salt, saturated fats, and carbohydrates, high in polyunsaturated oils, roughage, vitamins, and minerals. This diet, together with the Kung's physically active and relatively tension-free life-style, helps explains their low incidence of high blood pressure, hypertensive heart disease, high cholesterol, obesity, varicose veins, and stress-related diseases such as ulcers and colitis. The life expectancy of Kung adults is greater than that in many industrialized countries. On the other hand, the Kung are more vulnerable to infant mortality, malaria, and respiratory infections, as well as to a high death rate from accidents due to the absence of doctors and hospitals. Western scientists who observed Kung communities found that about one-tenth of the total population was over sixty years old, or roughly the same percentage as in those agricultural and industrialized societies with customary medical-care systems.

It is equally significant that the Kung are able to carry on their hunting and gathering with much less labor than is exacted today from workers in agricultural and industrial societies. The forty-hour week, which was won only after long and bitter struggle, would be considered inhuman by the Kung of both sexes. They devote fifteen to twenty hours a week to gathering and hunting, leaving the rest of the week free for resting, playing games, chatting, sharing the pipe, grooming each other, and visiting friends at nearby camps. Since the necessary food supplies can be obtained with a relatively small labor investment, young people are not required to work. Not until their mid-teens do girls join their mothers foraging, and boys their fathers hunting. At the campsite, work is shared along traditional gender lines, women being responsible for child care, cooking the vegetables and small game, serving the food, washing utensils, and cleaning the fireplaces, while men collect firewood, butcher the game, cook the meat, and make the tools.

The underlying communalism of Kung kinship society is evident in the sharing of property. Each Kung band collectively "owns" about twenty-five square miles of surrounding land, this being the maximum that is logistically manageable. If any band experiences a temporary food shortage, it is expected to ask permission to gather food in a neighboring tract. Permission usually is given, with the understanding that the favor will be reciprocated if the occasion arises. Perishable foods, whether meat or plants, are shared by all band members, but tools and clothes are the private property of the owner.

Communalism extends from property sharing to the Kung's carefully regulated social behavior. If a hunter, for instance, is exceptionally successful and returns repeatedly with much game, measures are taken to dampen any tendency toward conceit or any desire to lord it over others. "We refuse one who boasts," explains a band member, "for someday his pride will make him kill somebody. So we always speak of his meat as worthless ... 'you mean to say you have dragged us all the way out here to make us cart home your pile of bones.' ... In this way we cool his heart and make him gentle."[11] After a run of successful hunts, the rising star finds it politic to ward off possible envy or resentment by retiring into inactivity and enjoying the benefits of the reciprocal obligations he has accumulated. In this way, the Kung preserve band harmony by alternating periods of hunting and credit accumulation with periods of quiescence, when hearts can "cool" and hunters are made "gentle."

Finally, Kung social life is exceptionally rich and satisfying. Huts are so small that they serve only for sleeping. Fires burn in front of each hut door, and all doors face toward a large communal space. The emphasis then is entirely on the band's common social life. Individuals seek not privacy but companionship. Two-thirds of their waking hours are spent visiting or being visited by friends and relatives from other bands. An anthropologist observer notes that the Kung "must be among the most talkative people in the world." The talk is about the day's experiences hunting and gathering, about food distribution, gift giving, and much-savored gossip and scandal. Music and dancing are also important band activities, as are initiation rites accompanied by myths and legends passed down through the generations. This interweaving of art, religion, entertainment, and education constitutes the basis for band tradition and cultural continuity. "Their life is rich in human warmth and aesthetic experience," concludes an observer, "and offers an enviable balance of work and love, ritual and play."[12]

The Kung way of life is not only "enviable," but also inherently stable, or at least it was so until recent times. It is a society in equilibrium—an equilibrium that prevails not only between individuals, but also between those individuals and their environment. Basic needs are satisfied in a nonexploitative fashion. Personal conflicts of course abound, but not institutional ones. In fact, as anthropologist Stanley Diamond concludes about food-gathering societies in general, "revolutionary activity is, insofar as I am aware, unknown. It is probably safe to say that there has never been a revolution in a primitive society."[13]

Not only the concept of revolution but that of reform, too, is alien to such a society, and naturally so, since prehistoric peoples assumed that after the creation of themselves, their culture, and their habitat, equilibrium simply continued and was destined to continue. What need was there to criticize their culture or to try to change it? Parents trained their children to do what they themselves did. Education was a mechanism for preserving, not for altering their world.

Despite its past stability, Kung society today is fragile, if not rapidly disintegrating, as are other food-gathering societies throughout the world. Having existed for many millennia, all are now crumbling and face a bleak future. Inherently well balanced and self-perpetuating as long as they were left alone, these societies could not continue to exist in isolation once agriculture appeared about 10,000 B.C. The impact of agriculturists on food-gathering peoples is all too apparent today in southern Africa. The Kung have contact with neighboring Bantu farmers, whose mode of life they envy. They covet their domesticated animals, their ready supply of meat, milk, and vegetables, the colorful dresses of Bantu women, and above all, their seductive alcohol and tobacco. Therefore, the Kung take menial, poor-paying jobs in order to earn money to buy these desired goods.

Contacts with the nonfood-gathering world have created serious problems for the Kung. These include the coming of venereal and other diseases, as well as the contamination of the Kung's springs by the cattle and goats of the Bantu. The grazing of Bantu livestock has also had the effect of denuding the terrain of plants on which the Kung have depended, as well as frightening off their wild game. Not only are the Kung being reduced to the status of virtual beggars and hangers-on in Bantu villages, but even worse is their absorption into the military machine of South Africa. Today almost half the Kung adults are employed in one capacity or another at South African military installations. At the same time, the authorities have issued licenses for the opening of local liquor stores, so that the Kung can use their newfound wealth to purchase Johnnie Walker Scotch and cigarettes. What is happening to the Kung is also happening to the Eskimos in the Arctic, to the Native Americans on reservations in the United States and Canada, and to the aborigines in Australia.

Despite an antiquity that no other social system can remotely approach, all food-gathering societies have always faced the prob-

lem of what has been aptly defined as "the imminence of diminish-
ing returns."[14] After several weeks or possibly months in one location, food resources are depleted and a band must move on to a new
site. Hence, the constant nomadism, and the lack of any incentive
either to build up food reserves beyond a certain minimal point or
to construct substantial housing. Generally, only enough food is
collected to meet the needs of the moment, and the birth rate must
necessarily be kept low since too short an interval between births
would create unmanageable problems for the mother. Two infants
could neither be so easily breast-fed nor carried during foraging
expeditions or treks from camp to camp.

In fact, the unpleasant but inescapable task of population control
forced parents in many food-gathering societies to endure long periods of sexual abstinence after the birth of a baby. In others, an
infant born too soon after its sibling or together with a twin, might
be exposed to the elements, while old people might voluntarily end
their lives lest they become too much of a drain on food supplies or
too much of a burden on the community during migrations. Peter
Freuchen, who lived for decades with the Eskimos, has described
the poignant yet dignified suicide of an aged grandmother who could
no longer keep up with her band, and whose aching bones and
wheezing lungs made life a burden for herself and for her kin. Freuchen concludes that suicides are common when, as the Eskimos
express it, "life is heavier than death; [when] old men and women
are burdened with the memories of their youth, and can no longer
meet the demands of their reputation. . . . Fear of death is unknown
. . . they merely say that death can be either the end of it all or a
transition into something new, and that in either case there is nothing to fear."[15]

The end result has been an extraordinarily stable mode of life
with a built-in equilibrium, but also with a built-in Achilles' heel.
The population of any food-gathering society was bound to remain
sparse since far fewer of them could support themselves in a given
area than could food producers. Consequently, once agriculture
made its appearance, the food gatherers, unable to hold their own,
were pushed aside by sheer weight of numbers, an encounter between two ways of life made all the more unequal by the allure of
alcohol and nicotine, and by the material plenty of cultivated fields.

Equally lethal today has been the essential incompatibility between the cooperative, nonaggressive food-gathering way of life, and the competitive, acquisitive, consumer-oriented ethos of much of the twentieth-century world. "It is almost a contest of values," concludes an anthropologist analyzing the current situation on the Crow Indian Reservation in Montana. "The Crow believe in sharing wealth, and whites believe in accumulating wealth. . . . As a consequence the Crow have been paying a very heavy price for adhering to their traditional outlook."[16] That price includes an unemployment rate on the reservation of 85 percent, and a death rate from alcohol abuse eleven times higher than the national average. Analogous patterns prevail among native populations on all continents. As a result, while only ten thousand years ago, food gatherers made up 100 percent of the five million human beings who then inhabited the globe, today they number far less than 1 million out of a total world population of 5 billion.

LIFELINES

Although kinship societies are now unraveling, the fact remains that theirs constitutes virtually the totality of human experience in all but the past few millennia. It may be assumed, therefore, that they have left a legacy—strands of Ariadne's thread—that might serve as lifelines, providing meaning to the past, and guidance for the present and future. In fact, in at least four crucial areas, lifelines do extend from the experiences of those food-gathering ancestors all the way to our own lives.

ECOLOGY

A basic lifeline between prehistory and the present concerns ecology, or the interrelationship between humans and their home (*oikos*) on planet Earth. Anthropologists agree that Paleolithic food gatherers had a limited impact on their environment but disagree as to why this was so. Some maintain that it was neither the social organization nor the communal attitudes of kinship societies, but their lack of destructive technology that determined their minimal impact on their surroundings. Other anthropologists, more impressed by what they consider to be the food gatherers' reverential

30

attitudes toward their world, attribute that lack of impact to the nondestructive nature of kinship societies themselves.

Much evidence is available in support of either position. When American forest Indians gathered bark, for instance, they stripped it off only one side of the tree so that the tree would not be girded and killed. American Indians, however, also used their knowledge of fire to burn down whole forests in order to replace them with grassland that attracted more desirable game. Likewise, Peter Freuchen noted that both Indians and Eskimos "have the greatest respect and love for the animals they live on. They imbue them with souls, and it is their great concern that they shall feel good about being hunted and killed. Therefore, the animals play a big role in their folk stories."[17] Other anthropologists attribute this attitude not to reverence for their world but to the desire to make game more plentiful and easier to kill through magical means. They also note that when the ancestors of the Indians improved their technology with spear throwers and bows and arrows, they did not hesitate to hunt to extinction certain dangerous rival predators like the cave bear and the cave hyena, as well as the mammoth and the woolly rhinoceros, which were valued meat sources. All over the world, animals were exterminated by early humans: the moas of New Zealand, the giant lemurs and elephant birds of Madagascar, and the flightless geese of Hawaii.

Behind this possibly insoluble debate between anthropologists over the ways kinship societies related to their *oikos* lies a larger debate over human nature itself: are we "by nature" a rapacious, destructive species held back only by the limits of our technology, or have our "natures" for most of human history been relatively attuned to nature itself? What conclusions we draw about our possibilities in the present moment depend to a great extent on the position we consider most historically valid in this debate.

However, whatever the conscious or less than conscious motivations of Paleolithic humans, the fact remains that their small numbers and limited technology minimized their imprint on the environment. A recent study of the Semang, a group of two to three thousand nomadic forest foragers in Malaysia, reveals indiscriminate use of their river as a bath, laundry, toilet, fishing grounds, and source of drinking water. Likewise, they freely set fire to the surrounding forest in the course of their slash-and-burn cultivation, plant seeds in the fire-cleared ground, then resume their nomadic foraging, returning only to "harvest" whatever the animals have left in the unprotected plots.

Given this mode of life, the average per capita daily consumption of energy by the Semang does not exceed 5,000 kilocalories (40 percent from human labor and 60 percent from burning firewood), in contrast to the average 250,000 kilocalories expended by the typical U.S. citizen.[18] The scale of Semang actions is so small compared to that of our world that the results of what would be destructive acts for us may be quite the opposite in the Semang environment. The use of a river as a toilet by a small band results not, as we might expect, in pollution, but in the nourishment of a nutrient-poor tropical river while indiscriminate firing of the surrounding forest takes place on such a small scale that it has no appreciable effect on the Malaysian tropical rain forest as a whole.

Although it is indisputable that kinship people had "destructive" effects on their environment, it is also indisputable that they lived in a fundamentally different relationship to that environment and conceived of it differently than peoples of the historic period did. Whatever their destructive acts, they generally saw themselves as a living part of a living world that supported, encompassed, and enveloped them. As Peter Nabokov has pointed out, speaking of Native American peoples,

> most of them conceived of the earth and heavens as a cosmic temple. Many tribes could point to the exact cave or hill or lake from whence, according to their mythology, they as a people had first emerged from an ancient underworld. The surrounding trees, plants, seas, rivers, deserts, animals, and other forms of wildlife all figured as supernatural forces in their legends. Above was the canopy of Father Sky; below, the enduring body of Mother Earth.
>
> Their spiritual reference points were prominent environmental landmarks. In Arizona the Navajo universe was bounded by four sacred peaks. A hundred miles to the east, the San Juan Pueblo's world was circumscribed by four different sacred mountains. Everywhere tribes enshrined caves, springs, mesas, and lakes, leaving prayer sticks and food offerings for the spirits there.[19]

In conclusion, it should also be pointed out that, while all societies have always had a "destructive" or polluting impact simply by virtue of being human and using their technology to exploit the physical environment to satisfy their needs, the meaning of that "destructiveness" has not been the same in all times. The pollution of a Semang band practicing slash-and-burn agriculture is qualitatively—and conceptually—different from that of 5 billion humans deploying their high technology. The Semang have no reason to reflect on what effect they may be having on their river or their forest. Such "complacency" was understandable in premodern times but is vanishing as a viable option in a world where rivers, even oceans, can be turned into the equivalent of industrial sewers and waste-disposal dumps.

GENDER RELATIONS

The relationship between the sexes in food-gathering communities is of particular significance for us. Studies of some ninety food-gathering bands indicate that the status of women has been more equitable in kinship societies than in agricultural villages. The basic reason for this appears to lie in the fact that women contribute to the bands' food supplies at least as much as, and usually more than, male hunters do. Unlike village women, who are often confined to domestic quarters, food-gathering women are active band members, ranging far and wide in their foraging. Significantly enough, the degree of gender equality decreases in such bands in proportion to the importance of meat in the band's diet, large game meat invariably being the food provided by men. So sex equality is pronounced among Tanzania's Hadza, who consume little meat, while among the Eskimos, who eat only meat, women are treated largely as sex objects and have little control over their own fates.

The Eskimos, however, are the exception among food-gathering peoples. Among the Australian aborigines, for example, it has been noted that if a hunter returns to camp and finds his wife has not done her chores, he may berate or even try to beat her. She, however, is not likely to accept such punishment submissively; and if she, in turn, feels that her husband is lazy or fails to provide sufficient meat, she does not hesitate to attack him with both her tongue and her fighting stick.

The relative independence of food-gathering women has been

reported since Europeans first came into contact with such peoples centuries ago. European observers were usually surprised and shocked by the equality they enjoyed. The Jesuit missionary Paul le Jeune spent the winter of 1633–34 with a band of Naskapi Indians on the Labrador Peninsula of eastern Canada. "The women have great power here," le Jeune reported, and he urged the Indian men to assert themselves. "I told him that he was the master, and that in France women do not rule their husbands." Another Jesuit father noted, "The choice of plans, of undertakings, of journeys, of winterings, lies in nearly every instance in the hands of the housewife." Anthropologist Eleanor Leacock found that equality in sex relations still existed among the Naskapi when she studied them in 1950. "It was beautiful," she wrote, "to see the sense of group responsibility . . . and the sense of easy autonomy in relationships unburdened by centuries of training in deferential behavior by sex and status."[20]

Nonetheless, hunting-gathering peoples throughout the globe have divided labor responsibilities along sex lines, men being the hunters and women the gatherers. The explanation for this omnipresent division appears to lie in the biological fact that females are constrained from full-time hunting by the double burden of carrying the fetus during pregnancy and nursing the infant for at least two years. The periods during the life span of an adult female in a kinship society when she is neither pregnant nor nursing are relatively brief and sporadic. This makes it more practical and productive for females to concentrate on food gathering, an activity far less impeded by pregnancy and the nursing of infants. The division of responsibilities along sex lines explains why bands commonly were led by headmen rather than headwomen. The predominant role of males in the vital activities of hunting and defense-offense conferred upon them an aura that usually made them the generally recognized, albeit informal and transitory, band leaders.

If gender relations during the great preponderance of human history have been generally equitable, it may be assumed that the post-Paleolithic assumptions—so familiar in our own world—about an inherently "weaker" and "subordinate" sex are simply rationalizations justifying existing institutions and practices rather than scientifically established truths grounded in some sort of inherent female state of inferiority.

SOCIAL RELATIONS

Going beyond gender relations to overall band organization, we find that all surviving kinship bands have in common the distinctive characteristic of cooperation and communal sharing. This is due partly to nomadism, which makes individual accumulation of material possessions an encumbrance rather than an asset, partly to the fact that communal sharing gives a band a better chance for survival than competitive relationships do, and partly to the years in which an extended "family" can provide economic security to individual members who become incapacitated.

Yet whatever equality of wealth existed and exists in such bands, there is no such equality in prestige. Individual status varies substantially according to various criteria. Honor and respect customarily are accorded to those advanced in years, as well as those considered to possess supernatural powers. Similar high status is bestowed on individuals with outstanding personal qualities, such as skill in hunting or oratory, or exceptional generosity with material possessions or personal services. While a combination of such desirable traits may advance some individual to the rank of headman of the band, his authority is likely to remain minimal. It rests only on his personal qualities, not on any strict obligation to follow a leader or some decreed punishment for failure to do so. The headman is expected to continue making his own tools and to share in all work and communal obligations. He retains his position only so long as there is general satisfaction with his performance. Consequently, it is impossible for any individual to institutionalize his prestige or perpetuate his authority. As Paul le Jeune reported about the Indians among whom he lived in the 1630s: "They have reproached me a hundred times because we fear our Captains, while they laugh at and make sport of theirs. . . . [They] cannot endure in the least those who seem desirous of assuming superiority over the others; they place all virtue in a certain gentleness or apathy."[21] Typically, a report by a settler in Australia in 1867 strikes the same basic note on the egalitarian attitudes of the aborigines. "They do not understand exalted rank and, in fact, it is difficult to get into a blackfellow's head that one man is higher than another."[22]

Even though the distinctive communal and sharing characteristics of food-gathering bands persisted through millennia, they have proven fragile on contact with hierarchical or competitive societies. An example of this vulnerability is the case of an enterprising Kung man named Debe who set out to imitate the neighboring Bantu. He assembled a herd of goats and cattle, and was on his way to becoming a successful herder like his neighbors. Soon, however, relatives began visiting him when meat was scarce, and Debe, under heavy social pressure, had to slaughter his animals one after another to feed them. After several years in which his new enterprise and his traditional social obligations seemed in irreconcilable conflict, Debe sold or gave away his remaining animals and abandoned his effort to cross over to a different social order.

This fragility of kinship societies in the competitive modern world is recognized and emphasized by the Crow Indian Dale Old Horn, a graduate of the Massachusetts Institute of Technology and an instructor at Little Big Horn College in Crow Agency, Montana.

> Money and material wealth is not as meaningful to the Crow as kin ties and clan relationships. That fact is evident in Crow celebrations where individuals living on federal subsidies give away armloads of clothes, blankets or tools to friends and relatives.
>
> The Crow Indian child is taught that he is part of a harmonious circle of kin relations, clans and nature. The white child is taught that he is the center of the circle.
>
> The Crow believe in sharing wealth, and whites believe in accumulating wealth.
>
> It is a difference of view that results in poor aptitudes for business, and few defenses against whites who have mastered making a quick profit. It is nowhere more evident than in dealings over land.
>
> We have Indians here who are so broke they sell their land to non-Indians at rock-bottom prices just to get a little cash in their hands.[23]

Although the sharing behavior in kinship societies was the accepted human norm for millenia, it is important to note that it was not a mode of conduct that came either automatically or effortlessly. It was instead the end result of careful socialization during childhood. Anthropologists observing this socialization process in the Kung conclude that their infants, like all infants, are born with the capacity to be selfish as well as to share. Contradictory impulses are, however, channeled into socially acceptable forms that require more sharing and less personal accumulation than is customary in Western communities. Yet the tension between giving and taking is never resolved completely, as is evident in this not atypical complaint of an elderly man asking for a blanket: "All my life I've been giving, giving; today I am old and want something for myself."

As in the area of gender relations, where assumptions about a "second sex" for long were taken for granted, so in social relations individual acquisitiveness is often assumed to be a natural human norm. "A truly communal life is often dismissed as a utopian ideal," observes anthropologist Richard B. Lee, "to be endorsed in theory but unattainable in practice. But the evidence for foraging peoples tells us otherwise. A sharing way of life is not only possible but has actually existed in many parts of the world and over long periods of time."[24]

Food-gathering communities were unique not only for their relatively egalitarian social relationships, but also for their integration of work into daily life. Work was not a necessary evil, tolerated merely for the sake of sustaining life. Digging for roots, setting a trap, or chipping a stone tool were as much a part of daily routine as eating or telling stories or visiting a neighboring band. Work was not a means to an end, but rather both means and end. Modern workers hold jobs in order to earn money for subsistence and for the recreation necessary to "recharge the batteries," a phrase especially revealing in its implication of the enervating effect of daily work. The dichotomy between working and living that we take for granted would have been incomprehensible to Paleolithic band members.

Another unique feature of prehistoric work was its brevity. Studies of food-gathering bands on all continents reveal a common pattern of twelve to twenty hours of work per week. This was noted by the Jesuit Father Baird, who in 1616 expressed his astonishment at the ease with which the Micmac Indians of New Brunswick, Canada, obtained all the food they needed: "Never had Solomon his mansion better regulated and provided with food. . . . their days are all nothing but pastime. They are never in a hurry. Quite different from us, who can never do anything without hurry and worry."[25] Likewise today, Kung males are not expected to work before they marry and assume family obligations. Hence, the spectacle of able-bodied adolescents visiting neighboring camps or amusing themselves on their own. Their active hunting years are relatively brief since most retire in their fifties. In other words, at any one time about 40 percent of the Kung population contributes nothing to the common larder.

The ease and leisure of food-gathering societies when set against the toil and pressure of succeeding agricultural and industrial societies explains Marshall Sahlins's thought-provoking conclusion that "the amount of work per capita *increases* in proportion to technological advance, and the amount of leisure *decreases.*"[26]

WAR

Newspapers and television screens throughout the world flashed the exciting news in 1971 that a tribe of twenty-seven food-gathering people had been discovered living in total isolation on Mindanao Island in the Philippines. The striking and significant characteristic of these Tasaday, as they were called, was their complete lack of aggressiveness. They had no words for weapon, hostility, anger, or war. They gladly adopted the long Filipino knife, the bolo, because it was more efficient than their stone tools for gathering food and chopping their way through the surrounding jungle. But they showed no interest whatsoever in spears or bows and arrows because they were of no use for food gathering. As nonacquisitive as they were nonaggressive, they divided carefully and equally among all the members of their band whatever food they collected (yams, fruit, berries, flowers, fish, crabs, frogs).

The news about the "peaceful Tasaday" was welcomed by many because it seemed to undermine the popular theory that human beings are innately aggressive because of their genetic inheritance. According to this theory, our early vegetarian ancestors used their superior brains and tools to hunt and prey on other animals, thus acquiring a taste for flesh and becoming carnivores. Over the millennia, according to this school of thought, *Homo sapiens* became genetically programmed for aggressive behavior. Hence, the increasingly bloody conflicts throughout history culminating in the two World Wars and the holocausts of the twentieth century.

This idea of humanity as an inherently bloodthirsty species had been popularized and won a following in the 1960s. It is understandable, then, that among those who preferred to think of humans as an inherently peaceable species the news about the Tasaday should have been greeted enthusiastically. It was also noted that other food-gathering groups scattered throughout the world were equally nonaggressive. So the Tasaday and their like seemed to lead to an image of humans as innately peaceful and therefore not foredoomed to endless wars and ultimate self-destruction.

Unfortunately for this comforting hypothesis, not only was the veracity of the Tasaday story thrown into question by later press exposés, but at about the same time another band of thirty people, the Fenton, were discovered in New Guinea. These tribesmen were fierce warriors, continually fighting with their neighbors. Also in New Guinea were the Asmat headhunters who, according to a missionary living with them, devoted their lives to waging war. "All prestige, and therefore all authority, is ultimately derived from achievements in war. It is impossible to be a man of social standing without having captured a few heads. A bunch of skulls at the door post is a measure of status. . . . Successful headhunters enjoy many privileges: they are entitled to wear their ornaments as distinguishing marks; they can expect an extra portion of food . . . they need not exert themselves with heavy work; they are to be consulted in the meeting of men; they stand better chances with the women."[27]

Similar contradictions are to be found among food-gathering peoples on all continents. Among the American Indians, the Hopis and the Zunis raised their children for peaceful living, while the Cheyenne and Crow hailed military prowess as "the only road to distinction [for] the whole population, from cradle to grave."[28] Likewise, in Africa the Kung deplore aggressiveness and bravery, so that the heroes of their legends are animals that survive through trickery and deception rather than force. By contrast, the Boran pastoralists of East Africa spend their lives training to be warriors, and stress the importance of "valor and aggressive virility . . . stamina . . . bush skill . . . eye for ground and for cover."[29]

Where, then, does all this leave us regarding the nature of human nature? The overwhelming majority of scientists agree that humans, like other animals, have been self-serving throughout the millennia, and indeed had to be so in order to survive. For self-serving reasons, they evolved cooperative kinship societies during the Paleolithic millennia precisely because these were so well suited to the survival of the species under the conditions of those times. Human young, after all, are wholly dependent not for one year like monkeys or three to four years like apes, but for six to eight years. The survival of the human young during their long years of dependency was best assured by a system of cooperative base camps that assured the necessary food and protection.

Because in the Paleolithic period communal kinship society met certain crucial human needs, it persisted throughout the millennia. From this no one should assume either the absence of aggressive behavior or of conflicts in past or present-day food-gathering bands. Those who have studied the Kung, for instance, describe innumerable small clashes, often over women or simply because too many Kung had gathered in one area during the visiting season. But these clashes do not resemble anything that might be recognized as war. Instead, they usually involve a ceremonial display of hostility resulting in minimal damage or injury. Analyzing the nature of violence among the Yanomamo of the Amazon, Elman Service comments that its causes are "personal and familistic: usually the Yanomamo fight over women, sometimes over an insult, sometimes simply as bullying—to intimidate and prevent. When our teen-aged kids get in a fight for reasons like these we do not confuse it with organized warfare."[30]

Most revealing is the case of the Semai of Malaysia. In their native hills, they are completely nonviolent. They rarely show anger, will not kill an animal they have raised, and are known for their timidity. These same "gentle" tribesmen, when drafted into the Malaysian army and ordered to fight, become bloodthirsty killers reveling in butchery. On their return to Semai society, however, these soldiers again become as meek and averse to violence as their stay-at-home neighbors.

Such evidence leads psychologist Albert Bandura to conclude that "from the social learning perspective, human nature is characterized as a vast potentiality that can be fashioned by social influences into a variety of forms. . . . Aggression is not an inevitable or unchangeable aspect of man but a product of aggression-promoting conditions within a society."[31] This conclusion, supported by the evidence our own prehistory offers us, is basically encouraging. "Human nature" is neither pacific nor violent, neither cooperative nor predatory. It is rather, largely determined by "society" or "culture." But society and culture are made by humans and can be changed by humans. It follows that future societies and future humans will be determined not by preprogrammed genes for traits like acquisitiveness or aggression—as our recent historical experience sometimes appears to suggest—but by people who have the potential to be actors rather than pawns on the chessboard of history.

CHAPTER 2

TRIBUTARY SOCIETIES

Food gatherers were conscripts to civilization, not volunteers.
STANLEY DIAMOND

FROM KINSHIP SOCIETY TO
TRIBUTARY SOCIETY

The transition from kinship society to tributary society was a fateful turning point in human history. Compared to kinship societies with their sharing egalitarianism and warm communal bonds, tributary societies were impersonal and exploitative. They were based on tribute, in the form of goods, services, or money collected forcibly from a mass of cultivators and artisans for the support of a small ruling elite. It is understandable that the food gatherers involved in this transition were "conscripts" rather than volunteers, but how, then, could such a seemingly implausible transition have occurred in the first place?

The answer, on the basis of available evidence, appears to be demographic. Beginning about 10,000 B.C., population growth in the various regions of the globe reached such proportions that the traditional food-gathering technology no longer could meet the basic needs of the burgeoning communities. For millennia these food gatherers had limited the size of their bands by various conscious and less-than-conscious birth-control strategies. Breast-feeding, which suppresses ovulation, was commonly prolonged over several years with each infant, thus creating four- or five-year intervals between pregnancies. In addition, kinship peoples frequently had ritual taboos against intercourse not only after childbirth but also at other specified periods, which further helped to delay pregnancies. Hunter-gatherers also used various plants and herbs either as contraceptives or to induce abortions. If a new baby was born too soon after a previous birth, infanticide was commonly practiced because it was a hardship for women in such nomadic bands to breast-feed and carry two infants at once. These and other factors combined to

45

limit Paleolithic populations to an annual growth rate of 0.001 per thousand, which amounts to an increase for any given population of only 10 percent a century.

Even at this glacial rate of growth, our Paleolithic ancestors had increased to an estimated 10 million by 10,000 B.C. By that date, they had fanned out from their original home in the African savannas, had occupied all the continents, and were beginning to feel the pinch of overpopulation.

It would seem that with a mere 10 million people scattered over all of Africa, Asia, Europe, the Americas, and Australia, there would have been abundant room for all. In fact, this was not the case, because hunter-gatherers require an average of one square mile per person to subsist, and parts of all continents are too cold or too dry or too mountainous to support even that one person.

Population pressure became more acute with the passing of the last Ice Age. The melting of the glaciers caused approximately one-fifth of all previously available land to vanish below the surface of the rising oceans. This combination of decreasing land and increasing population meant that first in the Middle East about 10,000 B.C., and later in other regions, food gatherers were confronted with a disconcerting problem: their traditional way of life no longer adequately met their needs.

Band members who once had hived off to neighboring lands when food became scarce, now discovered that such lands were already occupied. This created a crisis that the kinship social system was incapable of meeting. Food-gathering bands had no built-in incentive to save or store for rainy days. Such days were effectively precluded by kinship sharing and by common access to nature's surrounding larder. In addition, constant nomadism made the accumulation of food or anything else logistically impractical. Hence, a way of life as relatively carefree and easygoing as outside observers invariably reported it to be became in certain areas less and less viable as the population grew beyond the one-person-per-one-square-mile average that hunting-gathering technology could accommodate.

The local breakdown of this social system that had endured so long and the contradictory impulses that ongoing events engendered must have been experienced as directly and forcefully ten thousand years ago as they were by the embryonic Kung entrepreneur, Debe, who attempted to raise livestock in our own time. It was in this way, then, that the kinship social system that had served human needs so well for millenia in the struggle against the vagaries of climate and the threat of predators now proved wanting in a struggle against population pressure.

The actual shift from kinship society to tributary society was gradual and prolonged. As hiving off into vacant lands became more and more difficult, the hunter-gatherers were forced to stay put and exploit local resources more efficiently. This they did in various ways. They harvested wild crops more carefully, increasing yields by weeding out unwanted plants, fencing off areas to keep out wild animals, and watering desired plants as needed. The Paiute Indians of the Owens Valley in California, to take a typical example, dammed small streams and dug irrigation canals, increasing the yield of wild rice that they themselves had not planted.

As the pressure for food mounted, women undoubtedly began planting the seeds of the most productive and nutritious wild plants they had previously simply gathered up each season. None of these developments required a prehistoric Archimedes shouting "Eureka" upon discovering the mechanics of agriculture. Women gatherers inevitably had to be very knowledgeable about the nature and behavior of plants in their habitats. Countless generations of food gatherers had acquired knowledge of how plants sprouted from seeds, how they grew better in certain types of soil than in others, and how much water and sunshine they needed.

In the early type of cultivation known as slash-and-burn agriculture, trees and undergrowth were cut down and put to the torch, leaving a residue of ashes that fertilized the soil. Seeds were then planted in the cleared nutrient-rich space, and the crops weeded and watered until harvesttime. Gardens were cultivated in this fashion until the soil became exhausted and the crop return on planting proved too small to justify the labor involved—a matter of one to eight years. These clearings would then be allowed to lie fallow for six to twenty years, while the land slowly returned to its original state as first tough grasses sprouted, then bushes and small trees, and finally larger trees. The land became ready for a new slash-and-burn cycle. This type of cultivation is still widespread, being practiced today by at least 300 million people in the Third World.

In this way, humans began to make their fateful transition from

food gathering to food producing, a transition repeatedly and inde-
pendently made all over the globe. The specific crops domesticated
depended on local environments and resources—millet in cold
northern China, rice in tropical Southeast Asia, wheat and barley
in the hot and dry Middle East, corn in the plateaus of Mesoamerica,
and potatoes in the high Andes. Of approximately 200,000 species
of flowering plants, only about 3,000 have been used extensively for
food, and of these, only about fifteen are of major importance: four
grasses (wheat, rice, maize, and sugar); six legumes or poor man's
meat (lentils, peas, vetches, beans, soybeans, and peanuts); and five
starchy staples (potatoes, sweet potatoes, yams, maniocs, and
bananas).

These plants were domesticated and their suitability to sustain
human populations reinforced through the centuries by selective
breeding for desired traits. Corn, for example, which originated from
the Mexican weed teosinte, had ears about the size of a thumbnail,
and miniscule kernels. Today the ears are a foot long, with many
rows of large kernels. In fact, corn is now so thoroughly domesticated that it cannot reproduce itself unless humans plant the
seeds. Some plants, like bananas and breadfruit have become so
domesticated that they no longer produce viable seeds at all, and
depend for their reproduction on humans to plant shoots from their
stalks.

The domestication of plants has not been a one-way process. If
humans domesticated plants, so did plants domesticate humans.
While hunter-gatherers subsisted on hundreds of plants and animals, enjoying a certain safety in the very variety of edibles available to them, agriculturists came to depend on the few crops they
grew, and so were far more vulnerable to disaster in case of drought,
flood, or pests. Humans were also quite literally domesticated by
their plants in that they had to give up their nomadic ways in order
to tend their crops and livestock. Soon the temporary camp gave
way to the village, where the overwhelming majority of human
beings henceforth were born, lived, and died.

Whatever its drawbacks, agriculture did solve the most critical
problem facing kinship societies—overpopulation. Even the early
slash-and-burn agriculture could support not one but an average of
ten people per square mile. This relief from population pressure,
however, proved short-lived because the transition to sedentary

village life quickly led to a population explosion. The new village life-style apparently triggered physiological changes in females that resulted in more frequent pregnancies.

It has long been recognized that postpartum menstruation, ovulation, and fertility are delayed in women who suckle their babies, so that breast-feeding acts as a natural birth spacer—nature's own contraceptive. In nomadic food-gathering societies such as the Kung, where breast milk is the sole nutrition for the infant, mothers are infertile for an average of four years. But when these Kung females give up foraging for sedentary village life, they reduce the length of breast-feeding substantially because they can wean their infants with grain meal and cow's milk. With this change, their period of lactational infertility shortens correspondingly. Continuing with their customary abstinence from any form of contraception, they now average five or six children per reproductive life span, as against the four or five during their earlier nomadic existence. The evidence from the Kung has been substantiated by further studies of other food-gathering peoples and, significantly enough, of North American Hutterite communities where mothers also feed their infants on demand and use no contraceptive procedures. All these studies disclose that continual breast-feeding induces prolonged periods of infertility and serves as a natural contraceptive that limits population growth.[1]

This physiological characteristic of the human species helps to explain why in the Middle East, where agriculture first began, the population jumped from less than 100,000 to over 3 million in 4,000 years, or a mere 160 generations. This precipitous growth promptly created another crisis of overpopulation that the new villages, based as they were on an antigrowth kinship societal system inherited from the food-gathering bands, could not handle. In other words, the technological revolution involved in the transformation from food gatherer to food producer was not immediately accompanied by a corresponding social revolution. Initially, the kinship system persisted everywhere with all inhabitants having free access to the fields surrounding each village, and the villagers therefore growing only enough to satisfy their immediate needs.

As studies of the Kuikuru Indians of the Amazon show, this can be done in an average of three-and-a-half hours a day, which is the full extent of their labor. A detailed account of how a kinship-based village functions is available in the following selections from the record kept by a Western observer of the work and nonwork stints of Bemba tribesmen in northern Rhodesia (now Zambia) during the month of September 1933.

September 1, 1933. Two gourds of beer ready, one drunk by old men, one by young men. A new baby born. Women gather from other villages to congratulate, and spend two or three days in the village. Women's garden work postponed during this time.

2nd. Old men go out to clear the bush. Young men sit at home finishing the sour dregs of the beer. More visits of neighbouring women to see the new baby. Few women go out to do garden work.

6th. Old and young men working by 6:30 A.M. and hard at it till 2 P.M. Two gourds of beer divided between old and young in the evening. Women working in their gardens normally.

7th. A buck shot by observer's party. Young men go out to fetch the meat. Women grind extra flour to eat with it. Two gourds of beer also made ready and drinking begins at 2 P.M. By 4 o'clock young men swaggering around the village, ready to quarrel, which they finally do. Dancing at night. Old women hilarious, and rebuked by their daughters for charging into a rough dance on the village square. Not enough beer for the younger women. They remain sober and express disapproval of the rest. No garden work done, except by old men.

8th. Every one off to their gardens in high spirits at 8 A.M. Back at 12 A.M. Young men sit in shelter and drink beer dregs for two hours.... Young girls go out on a miniature fish-poisoning expedition, but catch nothing.

17th. Great heat. Young men sit about in shelter all day, comb each other's hair, shave, and delouse each other. No relish available. Women too tired to cook.

19th. Nine men clear bush. One woman hoeing. Three women piling branches. Young women go fish-poisoning and catch one fish (about 2 lb.).

24th. Four gourds of beer divided between whole village. Sufficient for women as well as men. Beer-drinking lasts two days off and on.

30th. More beer. Four men clear bush.[2]

Assuredly, the social organization and the life-style of the early egalitarian villages had the same built-in brake on productivity as did the nomadic bands. Both afforded common access to the sources of livelihood, and therefore provided little incentive to produce more than was immediately needed. When the village population rose to the point where such an easygoing work schedule no longer met food needs, then social restructuring became unavoidable.

THE NATURE OF TRIBUTARY SOCIETIES

Just as population pressure forced the hunter-gatherers to exploit their local resources more intensively and in the process to develop agriculture, so now population pressure forced the early agriculturists to exploit their local resources more intensively and thus create tributary social organizations. This shift, which occurred first in the Middle East during the fourth millennium B.C., was repeated countless times throughout the world in the following centuries. The precise manner of the transition varied from region to region depending on the nature of the local physical environment, the types of crops grown, regional variants on the basic kinship social structures, and other factors.

In the Middle East, the hard-pressed farmers extended their cultivation from the uplands, where rainfall was usually sufficient for their crops, to the great river valleys of the Tigris-Euphrates and the Nile. There they found rich alluvial soil as well as date palms, reedy marshes swarming with wild fowl and game, and fish that provided protein and fat for the diet. A great handicap, however, was the searing heat and the sparse rainfall, which was much less than that in the surrounding hills. To obtain the necessary water, these pioneer farmers began to dig short canals from the river channels to their fields. The resulting crop yields were fabulous compared to those they had previously wrested from the stony hillsides. By the third millennium B.C., barley crops in the river valleys were eighty-six times larger than the quantity of seeds originally planted, and farmers were harvesting from their fields three times as much food as needed by their families.

Increased agricultural productivity provided the foundation—and also the inspiration—for the emergence of new crafts. One of the earliest of these was pottery, which began in the Middle East about ten thousand years ago and provided durable and clean storage containers for food and drink. In doing so, the new craft of pottery contributed fundamentally to the creation and maintenance of a communal surplus. Another early craft was metallurgy, espe-

51

cially useful in the river valleys where flint, the normal material used in the uplands for the making of tools and weapons, was unavailable. By 3000 B.C., the Bronze Age had begun, as alloys of arsenic and copper or tin and copper (both known as bronze) came into common use. By the same date, wind was being harnessed with square sails, that were initially used on boats in the Persian Gulf and on the Nile River. This represented the first successful use of inorganic force to provide motive power. Transportation on land also underwent an extraordinary change with the invention of the wheel, especially after about 3000 B.C., when the axle was fixed to the cart with the wheel able to turn freely, instead of the wheel and axle being rigidly fastened together. By that date, the wheel was being applied not only to land transportation but also to the making of pottery, which could from then on be turned out on a mass scale.

This combination of improved agriculture and new crafts was an explosive one, greatly increasing productivity, which in turn supported a growing number of villagers. While ten thousand years ago the world's population had been a little over 5 million, by the time of Christ it had jumped to an estimated 133 million. Teeming newcomers flowed out of the villages into new centers, or towns, which differed from the villages not only in size but also in social structure. No longer egalitarian units whose inhabitants all enjoyed access to the surrounding lands, the towns came to be dominated by a new ruling elite of kings, bureaucrats, priests, soldiers, and scribes, whose existence was only now made possible by the crop and craft surpluses generated by the cultivators and the artisans. Plato observed shrewdly that "every city is two cities, a city of the many poor and a city of the few rich; and these two cities are always at war."

The institutional mechanism this new elite used to maintain itself was the state, which possessed the authority and the personnel to skim off the surplus by levying "tribute" on the cultivators and artisans in the form of cash, produce, or labor service. Hence, the term "tributary society."

How and why did villagers surrender their egalitarian autonomy for the external authority and exactions of the state? This is a question much in dispute at present. The implausibility of the transition appears more plausible if it is not perceived as a black-and-white, one-way process in which the surplus of the cultivators and artisans was expropriated by an exploiting oligarchy that gave nothing in return. While the tributary state has often been described as an exploitative mechanism by which the "haves" exploited the "have nots" to preserve the stratified society under which they profited, the state also functioned as an integrative mechanism that provided protection, security, facilities for settling disputes, and access to sustenance in return for loyal acceptance of the status quo.

The ambiguous nature of the transition process to a tributary society is evident in the case of the priests—the first group of people to be freed from direct subsistence labor. They were successors to tribal shamans who practiced medical, magical, and religious arts to help and guide food-gathering bands, and who presided over modest shrines in the early villages. As those villages grew into towns, the shrines grew into temples, staffed by ecclesiastical hierarchies and sustained by elaborate theologies. This increasingly labyrinthine religious superstructure was supported by extensive endowments from the state, and varied contributions, voluntary and compulsory, from the faithful.

The priestly hierarchies provided in return important services for society. They were responsible for the invention of writing, which not only allowed them to keep a record of their numerous economic as well as religious activities, but to calculate the time of the annual floods, a matter crucial to every cultivator living in one of the great river valleys. They also kept the records essential for the proper functioning of the vast networks of dams and canals that spread through those river valleys. They were responsible for the vital decisions concerning the rationing of water and the construction and maintenance of irrigation works. They were also a key source of patronage and an important stimulus for the developing crafts, which at the outset produced as much for the temples as for any secular markets.

In the final analysis, the tributary state prevailed for five millennia from Mesopotamian times until the advent of capitalism in early modern times because it worked—because it satisfied the need for increased productivity to feed burgeoning village populations. Tributary societies were more productive than kinship societies precisely because they were exploitative class societies that levied and collected onerous taxes, rents, tithes, and labor services. To pay these exactions, the cultivators and artisans had to work infinitely harder than their predecessors had in the early egalitarian villages. How hard they had to work, how much they produced, and the marginal quality of life in a tributary society are evident in the following statistics for a forty-acre farm in Mecklenburg, northeastern Germany, during the fourteenth and fifteenth centuries. The farm produced an average of 10,200 pounds of grain each year, of which 3,400 pounds had to be set aside for seed for the following season's planting, and 2,800 pounds to feed four horses. Another 2,700 pounds had to be paid as dues to the lord, so only 1,300 pounds remained for the cultivator and his family. This amounted to a per capita daily ration of 1,600 calories, far below what we (or for that matter any Paleolithic food gatherers) would consider the minimum daily requirement. Beyond the long hours put in on farm work, the family had to spend extra hours growing vegetables and fruit in a garden, and raising livestock, poultry, and rabbits, simply to sustain life.[3]

THE GLOBAL VARIETY OF TRIBUTARY SOCIETIES

How different was the dawn-to-dusk drudgery of those German peasants and the easygoing existence of the Bemba villagers or the Kung hunter-gatherers. Exactly that difference explains why tributary societies prevailed and endured from the fourth millennium B.C. to modern times. During that long time span, many different types of tributary societies evolved on all continents, ranging from northern China to sub-Saharan Africa, and from Mesopotamia to the Aztec and Inca empires. Despite the chronological and geographic span, the Nile Valley cultivator in pharaonic Egypt and the Mecklenburg farmer in fifteenth-century Germany would have understood each other's daily struggle to sustain family life because both were tribute producers in essentially similar tributary societies. The underlying similarity, far outweighing either differences in crops produced or in political and cultural superstructures, was that both the Egyptian and the German cultivators lacked the free access to nature's bounty enjoyed by the Paleolithic food gatherers, and the free access to communal tribal fields enjoyed by early villagers.

Once such free access to a means of livelihood was severed, Paleolithic and Neolithic egalitarianism automatically gave way to inequitable tributary societies which, because of that very inequity, were far more productive and therefore prevailed until the emergence of modern capitalism.

The superior productivity of tributary societies also led to more complexly organized societies. The Bemba villages and Kung bands were socially egalitarian and culturally homogeneous. All members shared common knowledge, customs, and attitudes. In tributary societies, the collected tribute supported hierarchies that included monarchs, courtiers, administrative officials, judges, military leaders, priests, merchants, and scribes.

The tribute also supported a new "high" culture as opposed to the traditional "low" culture of the villages. This was a culture of the scribes who knew the mysterious art of writing, of the priests who knew the secrets of the heavens, of the artists who knew how to paint and carve, of architects who designed elaborate palaces, temples, and villas, and of poets, playwrights, historians, musicians, and philosophers, whose creations have survived to the present day. These two cultures, the high and the low, existed side by side, but ordinarily in more or less separate compartments. The high culture was to be found in the schools, temples, and palaces of the cities, the low culture in the ongoing life of the villages. The high culture was passed on from generation to generation among the educated in writing, the low culture by word of mouth among illiterate peasants. Anthropologists have focused their attention on the numerous low cultures of the world, while historians, depending mostly on written records, have focused on the high cultures, which they term "civilizations."

The first tributary society, or civilization, is generally believed to have emerged in Mesopotamia about 3500 B.C. It was followed by others: in Egypt about 3000 B.C.; in the Indus valley about 2500 B.C.; in China's Yellow River valley about 1500 B.C.; in Mesoamerica and the Andes about 500 B.C.; and in West Africa's Ghana about 700 A.D. These dates are no more than rough approximations, which are constantly being revised.

While the low and high cultures that developed on the various continents differed in details, they shared certain fundamentals. In the low cultures, for example, all peasants everywhere gradually accumulated a large body of information on the care of plants and animals. This corresponded to the vast lore possessed by hunter-gatherers concerning the plants and animals on which they subsisted. Peasants everywhere considered unceasing hard work a virtue—a characteristic that differentiated them from their food-gathering predecessors. They had little choice but to make a virtue out of a necessity imposed on them by the demands of tribute collectors and by the time-bound nature of stock raising and crop cultivation. Peasants also held in common a passion to own a plot of land and a few animals in order to gain some economic independence, no matter how marginal it might be. Hence, their dogged resistance in the past to the encroachments of secular or ecclesiastical landlords, and in the present to those of government collectives or agribusiness corporations. This emphasis on family independence was balanced, however, by the communal nature of village life. Peasants could commonly count on each other for aid and sympathy when needed, and were ready to participate cooperatively in house raisings, harvest festivals, religious celebrations, and other community affairs.

Similarly, the high cultures or civilizations of all continents shared certain common features. All were based on the collection of tribute in the form of cash, goods, or services. All possessed bureaucracies that collected tribute, dispensed justice, administered the country, and staffed the ecclesiastical structure. All had urban centers in which the ruling elite resided, and which served as the loci of institutionalized political authority. All maintained armed forces to repel external attacks and suppress internal subversion in case the bureaucracies and judiciaries failed to uphold state authority. All boasted monumental architecture—both imposing public buildings and the private residences of the elites. Finally, most of the civilizations were based on "sacred books" such as the Indian Vedas, the Buddhist Canon, the Islamic Koran, and the Christian Old and New Testaments. These texts dominated education—the young being required to memorize long passages—and were used as instruments of social control. Challenges to either religious doctrine or state authority were branded as crimes punishable in this world and the next. It was not coincidental that the "hells" that figured prominently in most of the sacred texts were in effect eternal concentration camps for any who dared resist the secular or ecclesiastical establishments.

Not all civilizations possessed all of the above characteristics. The Andean civilization of the Incas, for example, never developed a writing system. The Egyptian and Mayan civilizations lacked cities, at least as they are commonly defined. Some civilizations were based on slavery and slave labor; others were not. Some experienced lengthy periods of feudalism; others did not. Yet such individual differences were far outweighed by the cluster of common features that distinguished all tributary societies both from the kinship societies that preceded them and from the capitalist societies that followed.

Each of these civilizations developed through the centuries what might be defined as a distinctive "style." Very broadly, Indian civilization emphasized religion (as visitors can still discover in daily street scenes), Chinese civilization stressed propriety or suitable conduct for human beings (as expounded at length in the writings of Confucius), and Greek civilization centered on the individual (as reflected in the inscription at the principal religious shrine at Delphi, which read KNOW THYSELF, not FEAR THE GODS).

The civilizations of Eurasia crowd the pages of our world histories, while relatively little attention is paid to the high cultures of Africa and the Americas. However, when the early European explorers and freebooters from what was, in effect, a backwater of world civilization, first ventured overseas, they were as dazzled by the civilized wonders they stumbled upon in the Sudan and Mexico as by those they found in India and China. Conquistador Bernal Díaz, who participated in the conquest of the Aztec Empire, left memoirs that glow with descriptions of the creations of "the very skillful [Aztec] masters in cutting and polishing precious stones. . . . and the great masters in painting . . . and the wonderful sculptors." Díaz was particularly overawed by the Aztec capital, with its magnificent buildings, its aqueduct that "provided the whole town with sweet water," and its teeming, well-stocked marketplace. "Some of our men who had been at Constantinople and Rome, and traveled through the whole of Italy, said they never had seen such a marketplace of such large dimensions, or which was so well regulated, or so crowded with people as this one at Mexico."[4]

At about the time Díaz and the other conquistadors were destroying Amerindian civilizations, Western and Arab explorers and traders were sending back reports from Timbuktu and the coastal regions of West Africa that described flourishing agriculture, brisk trade, efficient imperial administration, and "the great store of doctors, judges, priests, and other learned men, that are bountifully maintained at the king's cost."[5] An English scholar has concluded that "allowing for the difference between the Moslem and the Christian intellectual climates, a citizen of 14th century Timbuktu would have found himself reasonably at home in 14th century Oxford. In the 16th century he still would have found many points in common between the two university cities."[6]

Despite the attainments of non-Eurasian civilizations, the fact remains that their legacy for the modern world has been relatively slight, because the course of modern world history has been so overwhelmingly dominated by Eurasia. One reason for this indisputable fact is that the Amerindian civilizations were strangled at infancy not just by conquistadors bearing swords and crosses, but by the even more deadly diseases—smallpox, measles, and malaria—European settlers and their African slaves unwittingly brought with them. Another reason for the predominant influence of Eurasia is that this great landmass is the heartland of planet Earth, comprising two-fifths of the landmass of the world, and including nine-tenths of the human race.

The peoples of Eurasia have enjoyed throughout history not only superiority in numbers and resources, but above all, the decisive advantage of accessibility. All the Eurasian civilizations—whether Chinese in the East, Indian in the South, Arab in the Southwest, or European in the West—were able through the millennia to interact and cross-fertilize. They exchanged their domesticated plants and animals, and also their inventions. The English philosopher Francis Bacon wrote in 1620 that three inventions—"printing, gunpowder and the magnet . . . changed the whole face and state of things throughout the world." All three were Chinese in origin, and all three were borrowed by the Europeans and used as basic instruments to power their own explosive chapter in global expansion. By contrast, all three, along with iron and the wheel, remained unknown to the isolated Amerindians until the appearance of Columbus.

The Eurasians in fact exchanged more than plants, animals, and technology; they also "traded" microparasitic infestations that gradually led to the development of Eurasian-wide immunities. Cholera and the bubonic plague, as well as cattle plagues, repeatedly devastated Eurasian lands, but eventually they lost their power to harm in their homeland. Let loose on the previously unaffected and therefore defenseless Amerindians and Australian aborigines, these diseases were at least as effective as firearms in clearing the way for the West's global hegemony. Early European settlers repeatedly reported coming upon Indian villages already cleared of their inhabitants by diseases that had spread ahead of them like wildfire, and with crops still unpicked in adjacent fields. The magnitude and horror of this holocaust is evident in the following 1586 report from Peru: "Influenza does not shine like the steel sword, but no Indian can dodge it. Tetanus and typhus kill more people than a thousand greyhounds with fiery eyes and foaming jaws. . . . smallpox annihilates more Indians than all the guns. The winds of pestilence are devastating these regions. Anyone they strike, they blow down: they devour the body, eat the eyes, close the throat. All smells of decay."[7]

A final point to be made about the world's high cultures or civilizations is the remarkable self-consciousness and resilience that has enabled them to survive, if not always flourish, to the present day. Outstanding in this regard are the Chinese, who have lived through innumerable military intrusions and cultural challenges without giving up their sense of Chineseness, their sense of themselves as the "Central Kingdom" around which the rest of the world revolves. Quite justifiably, the Chinese now can boast of being the world's oldest continuous civilization. Examples from China's long history demonstrate the strength of regional pride and self-consciousness, and hence the reality of civilizational durability.

In the seventh century, a Chinese Buddhist, Hsüan Tsang, visited the Nalanda monastery in India, which was then a center for the study of medicine, astronomy, mathematics, and magic. When he announced that it was time to go back home, the monks urged him to remain in Buddha's birthplace rather than return to a country of "unimportant barbarians . . . with narrow minds and profound coarseness. . . . That is why Buddha was not born there." Hsüan Tsang replied indignantly:

... in my country the magistrates are clothed with dignity, and the laws are everywhere respected. The emperor is virtuous and the subjects loyal, parents are loving and sons obedient, humanity and justice are highly esteemed, and old men and sages are held in honour. Moreover, how deep and mysterious is their knowledge; their wisdom equals that of spirits. They have taken the Heavens as their model, and they know how to calculate the movements of the Seven Luminaries; they have invented all kinds of instruments, fixed the seasons of the year. . . . How then can you say that Buddha did not go to my country because of its insignificance?[8]

Hsüan Tsang's response reflects how long-lasting a phenomenon intense pride in being Chinese and in things Chinese truly is. It explains why the many Chinese who converted to Buddhism between the fourth and eighth centuries turned it into something distinctively Chinese—something that coreligionists in Burma and Ceylon could hardly recognize as Buddhism.

The way medieval Chinese sinicized Buddhism has its counterpart in the way modern Chinese sinicized communism. In our century, millions of Chinese embraced communism and in 1949 won power under Mao Tse-tung's leadership, but the brand of communism they implemented was quite different from anything their Russian corevolutionists had developed. Mao's success derived in large part from his insistence on adapting Marxism to Chinese conditions and needs. In the course of his long revolutionary career, Mao repeatedly denounced abstract and doctrinaire Marxism and Marxists. "If a Chinese Communist who is part of the great Chinese people, bound to his people by his very flesh and blood, talks of Marxism apart from Chinese peculiarities, this Marxism is merely an empty abstraction. The Sinification of Marxism . . . becomes a problem that must be understood and solved by the whole Party without delay."[9]

Following Mao's death in 1976, the green light was given to private entrepreneurs, whose small cafés, fruit carts, blue-jeans stands, and flower stalls have prospered. Some are reportedly reaping profits far surpassing the salaries of high government officials. Yet in some cases these successful entrepreneurs are also experiencing a certain social ostracism. Just as Marxism was subject to sinification under Mao, so is private enterprise now under the re-

gime of Deng Xiaoping. A Shanghai newspaper reported early in
1986 that "entrepreneurs, even those with money, are having trouble
finding wives or husbands. The problem is that people are often
prejudiced against entrepreneurs. . . . They think some of the money
is earned dishonestly . . . people still consider entrepreneurs to be
of low social status."[10] Such popular sentiments appear to reflect not
so much a Communist disdain for capitalists as a traditional Confucian ethical disdain for merchants and money-makers. In Confucian
literature, the foundations of society are described as comprising
four classes—scholar-officials, farmers, artisans, and merchants,
with merchants always at the bottom of the social pyramid. Confucian sage Mencius reflected this anticommercial bias when he described one merchant as a "despicable fellow" because he was
always in search of "all the profit there was in the market. The
people all thought him despicable."[11]

The durability of tributary civilizations and the loyalty they have
commanded even in modern times is paradoxical in view of the
ambivalence with which they have always been regarded. Poets
and thinkers of all civilizations habitually have looked backward to
past ages with longing. They perceived their prehistoric ancestors
as "noble savages," untainted by the corrupting influence of civilized society. They fondly believed that long ago, "in the beginning,"
during the dawn of human history, there was paradise on earth. In
the Hindu epics, castes did not exist during that early dawn, and
humans enjoyed life in freedom and security. The eighth-century
B.C. Greek poet Hesiod wrote about an original Golden Age which,
with humanity's declining fortunes, was followed by Silver and Iron
Ages, ending wretchedly with what Hesiod considered to be the
deplorable era in which he lived.

One reason for this widespread millennia-old ambivalence about
tributary civilizations is that they were indeed an ambivalent phase
of human development. With their agricultural technology and their
tribute-based social organization, civilizations provided more goods
(often new) and sustained urban centers, with all their bustle and
commotion, their new interests, aspirations, and advances in the
arts of living. But this blossoming was accompanied by contradictions resulting in friction, tension, and insecurity.

A prime example of such inherent contradictions is the invention of writing. Apparently developed for the practical purpose of keeping accounts, writing turned out to have immense unforeseen consequences. It gave birth to literature, learning, holy scriptures and other forms of culture that could be transmitted and amplified from generation to generation. Out of all this activity emerged the creator, the self-conscious individual, who is the distinctive and precious product of civilization. The potential Aristotles, Michelangelos, or Einsteins who doubtless were born during the Paleolithic millennia had no chance of discovering their genius, much less of developing it. Tributary civilizations therefore were essential for nourishing the creativity of architects, sculptors, painters, and poets. The results of their creativity are visible and are enjoyed today in masterpieces such as the Parthenon and the cathedral at Chartres.

Yet the highest achievements of civilization involved the deepest contradictions. Writing stimulated creativity and intellectual growth, but it also became a tool of conservative elements interested more in preserving an advantageous status quo than in promoting free inquiry. They used writing to cultivate knowledge for decorative, sacred, and manipulative purposes. Hence, the "history" passed down through the ages is a distorted record as seen and interpreted by ruling elites. Mark Twain in *A Connecticut Yankee in King Arthur's Court* described forthrightly how traditional history has focused on the Reign of Terror during the French Revolution, while overlooking an infinitely greater reign of terror perpetrated during the entire course of French history.

There were two "Reigns of Terror," if we would but remember and consider it; the one wrought murder in hot passions, the other in heartless cold blood; the one lasted mere months, the other had lasted a thousand years; the one inflicted death upon a thousand persons, the other upon a hundred million; but our shudders are all for the "horrors" of the minor Terror, the momentary Terror, so to speak; whereas, what is the horror of swift death by the ax compared with lifelong death from hunger, cold, insult, cruelty, and heartbreak? ... A city cemetery could contain the coffins filled by that brief Terror which we have all been so diligently taught to shiver at and mourn over; but all France could hardly contain the coffins filled by that older and real Terror—that unspeakably bitter and awful Terror which none of us has been taught to see in its vastness or pity as it deserves.

A president of the American Historical Association, Lynn White, Jr., has analyzed the implication of such cultural skewing for our own times.

> From its beginnings until very recently written history has been a history of the upper classes by the upper classes and for the upper classes. . . . Yet we are only now beginning to realize that the inherited substance of our culture, despite its vast riches, is in many ways inadequate to our own times because it was originally cast in an obsolete aristocratic mold. We must write, and write from scratch the history of all mankind including the hitherto silent majority, and not merely that of the tiny focal fraction which dominated the rest.[12]

The millennia-old ambivalence in evaluating tributary civilizations simply reflects the ambivalence inherent in those historic social systems. They, and they alone, made possible the creation and accumulation of knowledge and the arts, and their transmission to successive generations. At the same time, this great leap forward achieved by humanity benefited for the most part the few rather than the many who, in the final analysis, bore the costs of high culture. The important point, so far as history and human lifelines are concerned, is that both material and cultural advances were made, and continue to be made, generating today as in the past, new problems along with new opportunities. Hence, the foreboding and jeremiads from our own latter-day Hesiods who look back with yearning to bygone golden ages.

That archcritic of civilization, Jean-Jacques Rousseau, proclaimed indignantly that man was born free but everywhere is in chains. But if Rousseau's contemporaries were weighed down by the chains of eighteenth-century Bourbon society, his prehistoric ancestors were weighed down by the chains of ancestral patterns that they accepted unquestioningly, never imagining the change, the self-development, and the free adaptation that became possible only with the complexity of civilization. The inescapable corollary of such "civilized" complexity is the insecurity and inequity which has ballooned to nearly intolerable proportions in our late-twentieth-century world. Insecurity and inequity, it must be emphasized, have been accompanied throughout the post-Paleolithic era by the growth of a sense of human potential—of an awareness of what has happened, what is happening, and what human beings could and should do about it. The total absence of revolutions and of reform movements during the Paleolithic millenia reflects not only social equity but constricted horizons.

LIFELINES

ECOLOGY

Hunter-gatherers left only a modest imprint on their physical environment in keeping with the modest level of their technology. With the coming of agriculture, however, the long-lasting equilibrium between humans and their *oikos* was permanently shattered. The surge of global productivity that both generated and was generated by the tributary system fed not just growing village populations but also the inhabitants of burgeoning new urban centers. It was not long before pioneer Mesopotamian towns of five to ten thousand people gave way to great imperial capitals with over a million inhabitants each. This population explosion was sustained for centuries by a continuing chain reaction in which population pressure triggered agricultural innovation, in turn stimulating further population growth.

The tributary state's most immediate and visible effect on the ecosystem lay in the destruction of the forests. As rapidly increasing populations felt the need for more arable land and grazing land, they naturally began to burn or cut down the surrounding forested areas. At the same time, their domesticated animals proved as destructive of forests as their domesticated plants. Goats caused surprisingly extensive damage as they browsed on shrubs, tree branches, and seedlings, eliminating all hope of forest regeneration in certain areas. Pigs rooted into the soil to get roots and seeds, especially acorns in oak forests. Sheep ate grass, roots and all, and their sharp hooves tore up the sod. Cattle, though not as directly destructive, required so much grass that farmers set fire to the forests simply to get the needed pasturage.

Finally, the forests suffered from the growing need for lumber to build houses, furniture, wagons, tools, and ships. As early as 2700 B.C., the Egyptians record the use of the famed cedars of Lebanon in constructing their fleets. A few centuries later, Sargon I, King of

Akkad, had the same cedars floated down the Euphrates for use in his Mesopotamian empire. Still later, the Phoenicians leveled whole cedar forests to build the ships with which they made themselves the leading traders in the Mediterranean region. In this fashion, century after century, the great cedar forests of Lebanon were steadily decimated until only about a dozen small groves survived, protected by walls that kept out both goats and lumbermen. That the trees in these enclosed areas still thrive is significant, disproving the common assumption that the deforestation throughout Mediterranean lands was the result not of human depredation, but of climatic change.

Contemporary observers who witnessed this deforestation process were fully aware of what was happening. Twenty-five centuries ago, Plato wrote the following analysis of how and why the green forested mountains of Attica were transformed into the sere brown landscape familiar today.

> Contemporary Attica may accurately be described as a mere relic of the original country. There has been a constant movement of soil away from the high ground and what remains is like the skeleton of a body emaciated by disease. All the rich soil has melted away, leaving a country of skin and bone. Originally the mountains of Attica were heavily forested. Fine trees produced timber suitable for roofing the largest buildings; the roofs hewn from this timber are still in existence. The country produced boundless feed for cattle. There are some mountains which had trees not so very long ago, that now have nothing but bee pastures. The annual rainfall was not lost, as it is now, through being allowed to run over the denuded surface to the sea. It was absorbed by the ground and stored . . . the drainage from the high ground was collected in this way and discharged into the hollows as springs and rivers with abundant flow and a wide territorial distribution. Shrines remain at dried up water sources as witness to this.[13]

Fully as devastating as deforestation were the vast irrigation networks that provided the economic foundation for many tributary civilizations. Just as the first cultivators could not have foreseen how their agricultural activities would lead to wholesale deforestation and erosion, so the first farmers who dug ditches to tap river water for their parched fields could not have foreseen how their irrigation activities would end up creating deserts rather than fertile fields.

At first, the irrigation works appeared to be an undiluted blessing. In the Babylonia of Hammurabi (eighteenth century B.C.), floodwaters were routed through irrigation canals lined with burnt bricks and connected by joints sealed with asphalt. The network laced the flat valley beds, and was steadily enlarged until at least ten thousand square miles were in crop, yielding enough produce to feed 20 million people.

However, this valuable economic structure was vulnerable because the Tigris-Euphrates rivers originated in the hills and mountains of Armenia, where overgrazing and deforestation had let loose an avalanche of silt containing salt and gypsum. Millions of tons of soil were swept down annually, causing the delta lands at the head of the Persian Gulf to advance southward into the Gulf an average of one hundred feet each year for the past five thousand years. Though a welcome addition to the supply of arable land, its value was far outweighed by the flood of silt that constantly clogged the canals and dams. Much labor power was needed to keep the vast irrigation system unobstructed and functioning properly. During times of foreign invasion or domestic disorder, these irrigation works tended to fall into disrepair, and then the river floodwaters would swamp the countryside instead of flowing smoothly through man-made channels to nourish crops. The excess water eventually evaporated under the broiling sun, leaving behind a residue of salt which, over time, transformed the once-fertile alluvial plains into the salt desert that comprises a large part of present-day Iraq.

Irrigation agriculture in the Nile Valley has not ended with such misfortune. Erosion in Ethiopia and Central Africa where the Nile originates has not been as severe as in Armenia's mountains. Consequently, the Nile carries only one-fifth as much silt as the Tigris-Euphrates, and therefore deposits only about one-twentieth of an inch of silt per year. This has been enough to replace the minerals lost with each harvest, but not enough to choke the irrigation canals. As a result, the Nile valley not only has retained its fertility through the millennia, but feeds many more people today than at any time in the past. Unfortunately, the Nile is the exception, and the Tigris-Euphrates the common pattern, whose calamitous results can be seen on all continents. "Civilized man," concludes one observer, "has marched across the face of the earth and left a desert in his footprints."[14]

66

In this manner, the balance between humans and their *oikos* changed fundamentally with the shift from food gathering to food producing. This, in turn, led to a corresponding change in attitude toward nature, which came to be regarded as a force to be tamed and harnessed to serve human needs. Greek philosophers, for instance, rejected traditional mythological and religious explanations of the natural world and sought understanding instead through the use of reason. For them the physical environment came to be seen not as the theater of the gods but as an object to be subjected to logical thought and analysis. The philosopher Protagoras best summed up this mental set with the statement that "man is the measure of all things," a concept even more strongly embedded in the Book of Genesis's injunction to humans to "have dominion over the fish of the sea, and over the fowl of the air, and over every living thing that moveth upon the earth."

The biblical mandate—and its equivalents in other tributary societies—was accepted and has been implemented to the present day. But as the twentieth century unfolds, with so many human achievements threatening to turn to ashes, a reassessment of the ecological experiences and traditional teachings of the great tributary civilizations and their successors in our time has begun, one that would have been inconceivable even half a century ago. The great civilizations of the past owed their rise and their splendor to technological advances that made possible a more effective use of the physical environment for human purposes. But when this use of the environment became abuse of the environment, then those civilizations invariably embarked on that self-destructive course that culminated in the "decline and fall" denouement around which so many world histories have been written. Mute evidence of this self-destructiveness is afforded by the ruins of cities, temples, and stadia amid deserts, and by the breached and silted remains of huge canals and reservoirs that no longer serve or protect any human population.

Today goats have been replaced by chain saws, and the dangers of silt and salt by those of chemical poisons and radioactive fallout. Little wonder, then, that the soil conservationist Walter C. Lowdermilk, after studying the global effects of environmental abuse, suggested that the traditional teaching that humans should enjoy priority over "every living thing" must be rejected. If God, he concluded, had foreseen the consequences of the misuse of our planet, he might have been inspired to give Moses another commandment, the eleventh:

> Thou shalt inherit the Holy Earth as a faithful steward, conserving its resources and productivity from generation to generation. Thou shalt safeguard thy fields from soil erosion, thy living waters from drying up, thy forests from desolation, and protect thy hills from overgrazing by thy herds, that thy descendants may have abundance forever. If any shall fail in this stewardship of the land, thy fruitful fields shall become sterile stony ground and wasting gullies, and thy descendants shall decrease and live in poverty or perish from off the face of the earth.[15]

GENDER RELATIONS

The advent of tributary societies altered relations between the sexes as fundamentally as it did relations between humans and their physical environment. When women gave up band life for village and town life, subordination and dependency ensued.

One reason was technological. The adoption of plow agriculture, large-scale irrigation systems, and new crafts such as metallurgy allowed men the possibility of cutting women out of full-scale participation in the new economy. The heavy labor involved in maintaining irrigation ditches, extracting tree stumps from fields, tending draft animals, or handling plows and other cumbersome equipment came to be seen as too difficult for women, who were considered hindered by physical constraints and by the need to birth and tend children. In this way, women lost control of the harvesting of the land, which had been the foundation of their equal status in kinship society, and instead found themselves spending more and more time at home, taking care of their children and husbands, and doing housework. No longer did they contribute equally to the food needs of the family. Instead, a distinction now arose between the "inside" work of women and the "outside" work of men, which came to be considered far more important, however essential the inside work might have been for family and for society. As "women's work" came to be regarded as less important, so women came to be viewed as the less important sex—the "second" sex.

With women more or less restricted to the home—or at least lesser work in the fields—men also gained dominance over the new state apparatus. They monopolized the assemblies, the courts, and the armies. They emerged with a virtual monopoly of economic power, political power, and military power. Women's role accordingly became a dependent and submissive one, that is, a powerless one, and they came to be called the "weaker sex." There were, of course, individual exceptions to this overall pattern of female subordination. Catherine the Great ruled the Russian Empire as ruthlessly and successfully as any tsar before or after her reign, and at the same time she at least held her own in diplomatic and military struggles with the contemporary rulers of the Hapsburg, Hohenzollern, and Ottoman empires. In China, Empress Wu Chao dethroned her son to become the only undisputed female ruler in Chinese history as well as a most capable and enlightened one. Yet such exceptional figures had no significant effect on the daily lives of their less exalted sisters, just as in our own time Evita Perón, Indira Gandhi, and Corazon Aquino have failed to improve significantly the daily lives of the women of Argentina, India, and the Philippines.

Women also lost status in tributary societies because a new emphasis on inheritable private property led men to go to great lengths to ensure that their personal wealth would pass on to their own children, whose paternity could not be left in doubt. As a result, among the elite, strict regulations and elaborate precautions were put in place to control female (though not male) sexual activity. These included chastity belts, bookkeeping devices to record the exact dates of sexual encounters, the castration of men who served as harem guardians for their rulers, and the widespread practice of clitoridectomy (cutting the clitoris out of young girls) which, by decreasing or eliminating sexual pleasure, was believed to be an effective method of keeping women from "straying" with men other than their husbands.

The precise manifestations of sexual inequality varied from one tributary civilization to another. In Mesopotamia, Hammurabi's Code in the seventeenth century B.C. described in clear legal terms the status of the husband as the undisputed head of the family. He was literally the owner of his wife and children, and could sell them to honor a debt. Adultery by the wife was punishable by death for both wife and paramour. For the husband, infidelity was not an issue because he had a legal right to a wife, a concubine, and slaves "for his desires" and to ensure descendants. In the illustrious democratic society of fifth-century Athens, women could not own property and had no political rights, including the right to vote. Demosthenes summarized the sexual inequity of classical Greece when he observed that there were three kinds of women: "mistresses [hetaera] for our enjoyment, concubines or prostitutes to serve our person, and wives for bearing our legitimate offspring."[16]

In China gender inequality was philosophically rationalized with the concept of two interacting elements, yin and yang. Yin, representing all things female, was seen as dark, weak, and passive, as against yang, the attribute of all things male, seen as bright, strong, and active. While both were considered necessary and complementary, nevertheless one was by nature passive and subordinate to the other.

Such was the ideological foundation for the inferior status of Chinese females, which began with birth itself, since baby girls more frequently suffered infanticide than did baby boys. Girls later were required to accept arranged marriages, after which they were brought within the realm of the husband's family and potentially subject to the tyranny of a mother-in-law. The most visible manifestation of gender inequity in Chinese society was the crippling of five-year old girls by the tight binding of their feet with yards of cloth until the arch was broken and the toes turned under. These bound feet, less than half the size of normal feet, were referred to as "golden lilies." They figured so prominently as objects of desire in poetry that no fewer than five divisions and eighteen types of deformed feet were identified. The cult of the "golden lilies" amounted to sexual psychopathology, with men finding the mincing gait and wriggling bodily movements that they forced on women highly erotic. As regards gender relations, foot binding literally made women mere sexual playthings. Their mobility was severely restricted, and they were rendered economically useless and completely dependent on men who could afford such human ornaments.

How much the status of women deteriorated with the transition from kinship to tributary society is reflected in the following conversation a Jesuit missionary reported having with a Naskapi Indian of Labrador in the seventeenth century. The Jesuit scolded the Naskapi for not acting as "the master," informing him that "in France women do not rule their husbands." The Jesuit related in his report the following exchange with the Naskapi: "I told him that it was not honorable for a woman to love anyone except her husband, and that this evil being among them [women's sexual freedom] he himself was not sure that his son, who was there present, was his son." The Naskapi replied: "Thou hast no sense. You French people love only your own children; but we love all the children of our tribe."[17]

SOCIAL RELATIONS

A paradox of the transition from kinship to tributary society was that as human productivity and wealth increased so did human malnutrition and starvation. One reason was the shift from the diversified Paleolithic diet of the hunter-gatherer to a limited village diet based on only a few domesticated grains. This left the majority of villagers vulnerable to the double jeopardy of malnutrition when crops were adequate, and starvation when they failed. But diet and weather alone could not have been responsible for the chronic malnutrition and starvation endemic to all tributary societies. Since these societies were far more productive than the preceding kinship societies, it follows that the chronic misery endured by villagers and townspeople was caused more by inequitable distribution as well as by insufficient production due to population growth.

Inequitable distribution was inherent in stratified tributary societies where the majority were denied access to the sources of livelihood as a social structure of landlords and landless, of haves and have-nots, came into existence. Such class stratification was characteristic of all tributary societies on all continents, since tributary societies by definition were made up of tribute collectors and tribute payers.

Full-scale exploitation of the have-nots did not necessarily flow instantly from the appearance of tributary social relations. The new elite class of specialists did provide services in administration, defense, religion, and crafts that benefited the entire community, so the exchange of services for tribute was initially of mutual, if not necessarily equitable, value. But everywhere such a mutual exchange sooner or later gave way to far more exploitative relationships as cultivators were forced to surrender their surplus without reasonable compensation. This invariably occurred—although at different periods in different regions—because all premodern civilizations remained relatively stagnant technologically after their initial breakthroughs in agriculture and the crafts. This left all tributary civilizations saddled with a level of productivity inadequate for their increasingly elaborate and costly superstructures. Inadequate productivity necessitated a ruthless extraction of all available surplus through the use of slavery, serfdom, debt bondage, rack renting, or other inventive forms of exploitation.

Studies of tributary societies, whether ancient, classical, or medieval, indicate that ruling elites made up only 1 to 2 percent of their total populations but appropriated not less than half of the total income. The agrarian and urban underclasses were often left with little more than what was necessary to ensure the biological reproduction of the labor force needed to support the state. Everywhere the privileged few lived in provocative luxury that contrasted sharply with the poverty and misery of the working many.

The sharpness of the contrast between the luxury at the top and the misery at the bottom varied from period to period. Since all civilizations were subject to the "rise and fall" syndrome, the degree of disorder, want, and exploitation naturally increased during the trough of this perennial cycle. Nevertheless, abundant contemporary testimony verifies the reality and magnitude of the social chasm ever-present in all tributary civilizations. As in the following passage from a first-century essay written in Han China, such testimony often also offers us some insight into the attitudes of elite observers toward the lower classes, who were customarily viewed as little better than animals.

> The gambler came upon a farmer clearing away weeds. He had a straw hat on his head and a hoe in his hand. His face was black, his hands and feet were covered with callouses, his skin was as rough as mulberry bark and his feet resembled bear's paws. He crouched in the fields, his sweat mixing with the mud. The gambler said to him, "You cultivate the fields in oppressive summer heat. Your back is encrusted with salt, your legs look like burnt stumps, your skin is like leather, that cannot be pierced by an awl. You hobble along on misshapen feet and painful legs. Shall I call you a plant or a tree? Yet you can move your body and limbs. Shall I call you a bird or a beast? Yet you possess a human face. What a fate to be born with such base qualities."[18]

Compare such an account with one on the life of the Chinese elite during the same period:

Opulent families lived in multi-storeyed houses, built with intersecting cross-beams and rafters that were richly carved and decorated on all visible surfaces. . . . In the inner rooms of the house the beds were carefully furnished with wooden fittings cut from the choicest timber; fine embroideries were hung up as drapes, and screens were set to overlap each other and ensure privacy. There was a shocking profusion of fine silk among the rich. . . . the wealthier classes wore choice furs of squirrel or fox, and wild duck plumes. . . . it was quite a common thing at a banquet to be regaled with one dish after another, with roasts and minced fish; kid, quails and oranges; pickles and other relishes. . . .

Naturally the rich families needed adequate means of transport up and down Ch'ang-an's streets [capital of Han Empire], and you could see their carriages drawn up in rows, gleaming in silver or gold and fitted with every sort of gadget. The horses themselves were neatly decked and shod, and caparisoned with breastplates and pendant jewelry. They were kept in check by means of gilt or painted bits, with golden or inlaid bridles. . . . With these extravagances there should be borne in mind the comparative cost of keeping the horses alive, as a single animal consumed as much grain as an ordinary family of six members.

There was no shortage of entertainment for the rich, who would amuse themselves looking at performing animals, tiger-fights and foreign girls. Musical performances were no longer restricted to special occasions such as folk festivals, and the tunes and dances were far more sophisticated than they had been in the past. Rich families now kept their own five-piece orchestras with bells and drums, and their house choirs. . . .[19]

Small wonder that popular complaints in imperial China often focused on the gross inequity between the gluttony of the rulers and the hunger of the ruled.

While the wine and the meat have spoiled behind the red doors [of rich households], on the road there are skeletons of those who died of exposure.

There is fat meat in your kitchen and there are well-fed horses in your stables, yet the people look hungry and in the outskirts of cities men drop dead from starvation.[20]

What a contrast between what the Chinese called this "man-eats-man" society, and a Native American kinship society, as described by a nineteenth-century anthropologist: "Hunger and

73

destitution could not exist at one end of an Indian village . . . while plenty prevailed elsewhere in the same village."[21]

The social inequity of tributary civilizations extended even to the dead. Archaeologists have discovered that disparities in grave offerings, while minimal in the early years of most tributary societies, increased as they became more complex and stratified. The great majority of graves were found to contain only a few pottery vessels or nothing at all, reflecting the poverty of commoners. Those of the well-to-do, however, might contain utensils, furniture, personal ornaments made of precious metals, and a myriad of other manifestations of conspicuous consumption. Royal tombs, in particular, stood out above all the others with their lavish accoutrements that might include not only beautifully wrought ornaments and weapons but also numerous personal attendants—men-at-arms, harem ladies, eunuchs, musicians, and general servants—all sacrificed so that they might serve the royal occupant in an afterlife while continuing to reflect his affluence and majesty here on earth. In one now world-famous example, the first great unifying emperor of China, Chin Shih Huang-ti, surrounded his tomb with an army of 7,500 terra cotta warriors, each life-sized with individually modeled heads representative of the ethnic groups under the emperor's rule two thousand years ago. Significantly enough, the senior members of the Bureau of Antiques Administration of Shanxi province (where this enormous mausoleum is located) were disgraced as reactionaries during the Cultural Revolution because they allegedly showed more interest in royal tombs than in the lives of poor peasants—thus reflecting the reality and vitality of class differentiation through the millennia.

Even the skeletons themselves reflect the harsh inequity of life in tributary societies. A study of skeletons at Tikal, Guatemala, in the first millennium A.D. disclosed that, while the average Mayan male was a mere five feet one inch in height, those few given elaborate burials had an average stature of five feet seven inches. Their bones also were substantially more robust and their life span clearly had been a longer one. Archeologists assume that a "nutritional advantage" allowed the Mayan elite to realize their full potential in stature and life expectancy.[22] It might be added that such skeletal differentiation on the basis of class appears to persist to the present. Britain's Health Ministry reported in December 1984 that, "In almost

74

every age group, people from households headed by a manual worker were shorter, on average, than people from non-worker households."[23]

If tributary societies were so exploitative, why were they tolerated by countless generations of people whose labor and support made possible their survival? The answer is to be found in a complex combination of institutional coercion, psychological manipulation, and direct military force. Even so, support from below was by no means a given in any tributary civilization. Indeed, the history of tributary societies was punctuated by recurring outbreaks of violent resistance, ranging from isolated local insurrections to mass jacqueries that threatened kingdoms and empires.

The full story of such resistance from below will never be known because, as the Zulu proverb reminds us, "The voice of the poor is not audible."[24] Since historical records are basically the records of a literate elite, the voice of the illiterate multitudes has been muted throughout the millennia. Even so, those records do yield ample evidence of recurring uprisings, invariably crushed with relentless severity.

The Roman Empire was racked by slave revolts in 419, 198, 196, 185, 139, 132, 104, and 73 B.C. The last of these, led by the Thracian gladiator slave Spartacus, was the most formidable. Three Roman armies were routed before the slaves were finally defeated. Six thousand of them were then crucified on crosses lined up for miles along the Appian Way, where their death agonies were watched by wealthy Romans on the way to their country villas. A few decades later, a similar drama was enacted at the other end of Eurasia, in the Chinese Empire. From 18 to 25 A.D., peasant rebels known as the "Red Eyebrows" overran the lower basin of the Yellow River and captured China's imperial capital. Immense casualties were suffered by both sides before the revolt was ferociously suppressed and the Han dynasty restored.

Such outbreaks continued into modern times. No fewer than 1,467 peasant outbreaks have been identified between 1801 and 1861 in tsarist Russia alone. The passions aroused by these eruptions are illustrated by the 1773–74 jacquerie led by the Don cossack, Emelian Pugachev, who issued a manifesto emancipating Russian serfs from all obligations to lord and state, and giving them possession of "the fields, forests, meadows, fisheries, and saltpans without cost."

Pugachev further commanded the serfs to seize and punish the nobles, whom he called "the disturbers of the empire and the despoilers of the peasants," and he promised that "after the extermination of these criminal nobles, everyone will be able to enjoy tranquility and a peaceful life that will endure till the end of time."[25]

The ruler of Russia at the time was Catherine the Great, who, despite her well-publicized claim to being an "enlightened" monarch, nevertheless issued a countermanifesto making clear her abhorrence not only of Pugachev's sedition but also of any acts or thoughts that challenged the divinely ordained rule of herself, of her fellow monarchs, and of the landowning nobility. "There is not a man deserving of the Russian name, who does not hold in abomination the odious and insolent lie by which Pugachev fancies himself able to seduce and to deceive persons of a simple and credulous disposition, by promising to free them from the bonds of submission, and obedience to their sovereign, as if the Creator of the universe had established human societies in such a manner as that they can subsist without an intermediate authority between the sovereign and the people."[26]

Catherine prevailed. Pugachev was defeated and executed, and serfdom continued for another century to dominate the lives of the great majority of the Russian people, who remained chattels on the nobles' estates. How deeply this unconscionable yet enduring social scourge affected sensitive Russian intellectuals is evident in the following reflections of the nobleman Nicolai Ivanovich Turgenev, who participated in the unsuccessful Decembrist uprising of 1825 and was sentenced to death *in absentia:* "There are some ideas which seize a man from the beginning of his life, which never leave him, and which end by absorbing, in one way or other, his entire existence. . . . From my childhood I felt a pronounced repulsion for serfdom; I instinctively sensed all the injustices in it. . . . The sympathies of my youth became convictions when I was able to examine the question, when I was able to estimate the enormous evil that serfdom produced, first for the serfs, then and above all for the serf-owners whom it degraded even more than the serfs, and finally for the state and for humanity, whom it dishonored."[27]

Despite the bloodshed and suffering endured in the course of these countless uprisings, all were suppressed, at least to a degree sufficient to enable the tributary societies to endure through the millennia, albeit sometimes with new faces in control of the state. The most obvious reason was that the raw power of the state was more effectively organized and could, in the long run, be more effectively deployed than that of the rebels. By definition, the state was the mechanism for mobilizing and projecting power through such state organs as the military, the militia, the police, the bureaucracy, and the judiciary. Buttressed by the revenues accumulated by its tax collectors, by the organized might wielded by its armed forces, and by the services rendered by its administrators and judges, the tributary state proved, in the end, capable of suppressing—or in rare instances, absorbing—any rebellious force it faced up until modern times.

As important as state power in preserving tributary societies was the massive indoctrination apparatus designed to persuade members of a given society that the status quo was not only legitimate and desirable, but God-given and irresistible; that any other way of organizing society was inconceivable, not to say utterly blasphemous. The Egyptian pharaohs, for instance, were regarded not only as the rulers of their country but also as "living gods," a coupling of divine and secular authority that left little space available even for imagining new ways of opposing the status quo, let alone organizing a different future. Equally powerful was the Hindu doctrine of karma, which held that one's status in life was dictated by one's deeds in previous existences. What opportunity could there be for individual assertiveness or the concept of changing your situation in life if present status was determined by past actions, and if the only hope for an improved status in a future life rested on dutiful observance of stipulated caste duties, regardless of how onerous they might be.

Generations of such mind conditioning inevitably left an indelible imprint. Aristotle stated the matter forthrightly in his *Politics:* "Some men are by nature free, and others slaves, and for these latter slavery is both expedient and right." Such ruling-class ideology was accepted and internalized by the overwhelming majority of subjects. In the words of Dostoyevsky's Grand Inquisitor, the millennia of tributary civilizations were the millennia of "miracle, mystery, and authority," when "the universal and everlasting craving of humanity was to find someone to worship." Tributary civilizations managed to monopolize the political imagination in such a way that alternative modes of social organization became almost impossible to imagine, or at least so vague and utopian (as in peasant millenarian movements) that they proved irrelevant to the real world.

In that real world of tributary societies, life came to be identified with work. One could even say that it was in such societies that the very concept of "work" first arose in a way that was unimaginable to kinship peoples. For one thing, the amount of work extracted from the cultivators who made up the vast majority in such societies increased sharply. In this way, paradoxically, each technological advance came to lead not to lighter but to heavier human work loads.

One reason for this lay in the new social organization of agriculture, which forced cultivators to work far harder simply to produce the surplus necessary for the support of a ruling elite. A second reason lay in the nature of the new agricultural technology, which required more labor than the one that had preceded it. No extra work load materialized as long as cultivators practiced only slash-and-burn agriculture, scattering seeds directly among the ashes in fire-cleared plots, or planting tubers of root crops with digging sticks. There was no need for fertilizers or for permanent irrigation networks as long as enough land was available to allow fields to revert to a wild state after a few seasons of crops. This type of cultivation actually required an average of only 500 to 650 hours a year of labor per person, somewhat less than that of food-gathering peoples.

As population increased, however, slash-and-burn agriculture was replaced by either plow or irrigation agriculture. Both were far more productive, but also far more labor intensive, requiring as they did terracing, or the building of irrigation works, or fertilizing, haying, and the feeding and care of draft animals. The easy life-style typified by the Bemba was replaced by grueling, dawn-to-dusk work routines commonly associated with village life. Time now became an important, even a pressing matter. Animals had to be fed and milked on schedule, crops planted or harvested in consonance with seasonal or weather demands. Peasants felt constrained as never before "to make hay while the sun shines."

Work in tributary societies changed fundamentally in quality as well as in quantity. Work became labor, separate from rather than an integral component of life, an unavoidable affliction necessary to sustain life. The nature of this qualitative change is evident if the Bemba life-style, in which all villagers participate voluntarily and spontaneously in communal work activities, is contrasted with the pronouncements of Plato about preordained work obligations. Each person, as Plato put it in his *Republic*, is destined to perform the work "for which he was by nature fitted," and within the class to which he belonged. At that occupation, Plato added, the citizen was "to continue working all his life long, and at no other ... in our State, human nature is not two-fold or manifold, for one man plays one part only." Plato's description reflects a two-fold shift in the nature of work: from voluntarism to compulsion; and from multiple skills learned as part of a growing-up process to single-occupation specialization enforced throughout life.

The subordination and exploitation of cultivators and artisans is strikingly illustrated in this third-millennium B.C. admonition by an Egyptian father to his son departing for school.

> Put writing in your heart that you may protect yourself from hard labor of any kind and be a magistrate of high repute. The scribe is released from manual tasks; it is he who commands. . . . Do you not hold the scribe's palette? That is what makes the difference between you and the man who handles an oar.
>
> I have seen the metal-worker at his task at the mouth of his furnace, with fingers like a crocodile's. He stank worse than fish-spawn. . . . The stonesmason finds his work in every kind of hard stone. When he has finished his labors his arms are worn out, and he sleeps all doubled up until sunrise. His knees and spine are broken. . . . The barber shaves from morning till night, he never sits down except to meals. He hurries from house to house looking for business. He wears out his arms to fill his stomach, like bees eating their own honey. . . . The farmer wears the same clothes for all times. His voice is as raucous as a crow's. His fingers are always busy, his arms are dried up by the wind. He takes his rest—when he does get any rest—in the mud. If he's in good health he shares good health with the beasts; if he is ill his bed is the bare earth in the middle of his beasts. . . .
>
> Apply your heart to learning. In truth there is nothing that can compare with it. If you have profited by a single day at school it is a gain for eternity.[26]

The hardships of daily life for those at the bottom of the social pyramid were tempered by holidays, celebrations, and religious observances. The human species by its very nature is *Homo festivus* as well as *Homo sapiens*. All cultures have their element of festivity. The Hindus revel at Holi. The Moslems feast after the long fast of Ramadan. Christians celebrate Christmas and Easter, as well as the sacraments—weddings, christenings, and burials. It is estimated that a medieval European peasant enjoyed an average of 115 holidays each year (in addition to the fifty-two Sundays).[29] Yet this same peasant undoubtedly felt both poor and victimized. A fourteenth-century English peasant expressed very pointedly this sense of class inequity. "Their [the lords'] satiety was our famine; their merriment was our wretchedness; their jousts and tournaments were our torments."[30]

This typical peasant attitude raises a question about the very nature and meaning of poverty. When it came to material possessions, after all, kinship societies inevitably lagged far behind tributary civilizations. Many accounts have been left of the tremendous variety and volume of goods displayed in the marketplaces of the great urban centers. When Marco Polo visited Hangchow in the late thirteenth century, he reported enthusiastically that this "most splendid city in the world" possessed "ten principal squares or market places, besides which numberless shops run along the streets. These squares are each half a mile in length. . . . capacious warehouses, built of stone, to accommodate the merchants from India and other countries. . . . there is an assemblage of from 40,000 to 50,000 persons, who bring for sale every desirable article of provision. There appears abundance of all kinds of game, roebucks, stags, fallow-deer, hares and rabbits, with partridges, pheasants, francolins, quails, common'fowls, capons, ducks and geese almost innumerable. . . . also cattle for food, such as oxen, calves, kids and lambs. . . . a great variety of herbs and fruits. . . . From the sea, twenty-five miles distant, a vast supply of fish. . . . and imported articles are sold, as spices, drugs, toys, and pearls."[31]

Equally striking was conquistador Bernal Díaz's account of what he saw in the market of the Aztec imperial capital, Tenochtitlán: "The lake was crowded with canoes which were bringing provisions, manufactures, and other merchandize to the city. . . . in this immense market we were perfectly astonished at the vast numbers of people, the profusion of merchandize exposed for sale. . . . I wish I had completed the enumeration of all this profusion. . . . the variety was so great that it would occupy more space than I can well spare to note them down in."[32]

Such affluence, even glut, would have proven far beyond the comprehension of food-gathering peoples. But should they be considered poor if the few articles they did possess quite adequately satisfied their physical needs? Marshall Sahlins has noted that poverty does not necessarily mean simply a paucity of personal assets. "The world's most primitive people have few possessions, *but they are not poor.* Poverty is not a small amount of goods, nor is it just a relation between means and ends; above all it is a relation between people. Poverty is a social status. As such it is the invention of civilization."[33]

Poverty indeed was invariably present in stratified tributary societies because it was of two varieties. There was the literal poverty of the have-nots, whose families suffered from chronic want, hunger, and starvation. There was also the poverty of those who, like the fourteenth-century English peasant, felt deprived and poor because others had so much more and flaunted their affluence. In a stratified society, regardless of that society's level of productivity and average personal income, there were always those who felt themselves have-nots simply because of such concepts and practices as private property, personal accumulation, family inheritance, and infinite desires as distinct from needs.

The poverty-free kinship societies of Paleolithic times hold an understandable appeal for some of us living as we do in an era of unprecedented productivity, an era in which unprecedented desires have been stimulated by a boundless consumerism. Yet the fate of the Kung today indicates that "poverty-free" societies could exist successfully only in an unstratified world where parents were able, without the competition of rival models, to teach their children to practice sharing and cooperation.

WAR

Tributary societies changed the nature of war as fundamentally as they did the basis for social relations. Kinship societies had no standing armies and no military specialists. Instead, ad hoc leaders emerged in a band by virtue of their individual prowess. Their command, accepted only for specific emergencies, was inherently personal and ephemeral. What they lacked was the support of any institutionalized authority outside themselves, or the backing of any sort of permanent professional cadre of soldiers. Nomadic bands also never had the resources for prolonged fighting. Since hunters were the source of a band's meat supply, they could not be spared for protracted warfare. Nor could a small band support either military specialists whose lives would be devoted to planning and waging war, or the warriors needed for even the tiniest standing army.

All this changed drastically with the organization of tributary societies. Large standing armies headed by professional warrior castes appeared for the first time. It was, of course, the tribute siphoned off peasants and artisans that provided the resources needed for military equipment, transport facilities, food supplies, fortifications, and all the other prerequisites for prolonged campaigning. As a result of this institutional transformation, warfare also was transformed from a series of episodic personal duels, raids, and skirmishes to the mass activity that has enmeshed entire societies and bedeviled the human species for the past several thousand years.

Tributary societies provided not just the resources but the incentives necessary for the new mass warfare to become commonplace. The developing crafts crucial to the wealth and well-being of the great urban centers depended on raw materials often imported from elsewhere. Since the Mesopotamian lowlands, to take an example, possessed neither metals nor quality timber, copper was brought in from Oman, silver and lead from the Taurus Mountains, and timber from the cedar forests of Lebanon. To pay for these imports, Mesopotamian rulers either had to ensure that the various crafts expanded production enough to furnish exports in exchange, or they had to take the sources of the needed raw materials by force. King Sargon opted for this alternative when he conquered an empire extending "from the Lower to the Upper Sea"—from the Persian Gulf to the Mediterranean. A contemporary poem, "The King of Battle," describes how Sargon led his forces through unknown mountain passes, and "did not sleep" in his campaigns for mastery of trade routes. "Brisk activity dominated the wharf where the ships are docked; all lands live in peace, their inhabitants prosperous . . . without hindrance ships bring their merchandise to Sumer."[34]

It was not only the aggressiveness of tributary societies seeking wealth, but also their need to defend themselves against the ceaseless attacks of surrounding nomadic peoples that led to large-scale institutionalized warfare. The rich civilizations of the Nile, Tigris-Euphrates, Indus, and Yellow River valleys were irresistible magnets for these nomads who inhabited the steppes and deserts between the river valleys. Abundant crops, barns swollen with grain, ornate palaces and temples, and the myriad commodities of the bazaar merchants and artisans, all drew the comparatively poor and hungering nomads.

These nomads first appeared in open grasslands where rainfall was inadequate for agriculture, and local inhabitants had no choice but to depend for their livelihood on domesticated animals rather than domesticated plants. In the vast steppe and desert areas from the Sahara to Manchuria, pastoral nomads had for millennia lived off their herds of cattle, horses, camels, sheep, and goats. Until the second millennium B.C., however, they lacked the military power to seriously threaten the centers of civilization. This imbalance of power in favor of tributary societies was only upset when these "barbarians" (as settled tributary societies thought of them) learned how to domesticate the horse and how to smelt iron. With these two achievements, the barbarian nomads became a formidable military force, possessing superior mobility and iron weapons as sophisticated and plentiful as those of the standing armies guarding the urban centers.

With their horses and their iron weapons, the nomads periodically launched invasions that overwhelmed all the centers of civilization between China on the eastern tip of Eurasia and Europe on the western tip. The earliest tributary civilizations were inundated by two waves of nomadic onslaught. The first of these, between 1700 and 1500 B.C., consisted of invaders armed with bronze weapons and riding horse-drawn chariots. The second wave of invaders, between 1200 and 1100 B.C., came on horseback and fought with iron weapons. Later classical civilizations were also uprooted by barbarian onslaughts, such as those of the Germans and Huns in the fourth and fifth centuries A.D. In medieval times, Moslems emerged from the Arabian deserts to conquer the Middle East and much of the settled Mediterranean world. Most fearsome of all, the Mongols burst out of the Central Asian steppes in the thirteenth century and conquered a vast empire stretching from the Baltic Sea to Southeast Asia, and from the Adriatic Sea to Korea.

Invasions and empires of such continental proportions required enormous military establishments. The Chinese, for instance, constructed their fabled Great Wall extending for 1,400 miles from Inner Mongolia to southern Manchuria. They manned this gigantic rampart with 300,000 worker-soldiers who were entrusted with the double duty of maintaining the wall and related fortifications, and also guarding against incursions by the nomads from the north. India's famed Emperor Asoka (273–232 B.C.) maintained a permanent army of 700,000 men supplied with 9,000 elephants and 10,000 chariots. The frontiers of the Roman Empire at its height in the third century A.D. were defended by Roman legions totaling 400,000 men, behind extensive fortifications such as Hadrian's Wall on the Scottish border, and connected their supply sources by a network of excellent roads that also enhanced their mobility.

The mobilization of such enormous material and manpower resources reduced to insignificance the campaigns of that pioneer Mesopotamian empire builder, Sargon, let alone the trivial aggressive activities of food-gathering bands. In the same region as Sargon's early empire, the Assyrian King Sennacherib conquered Babylon some sixteen centuries later. He left this account of how he dealt with it: "I levelled the city and its houses from the foundations to the top, I destroyed them and consumed them with fire. I tore down and removed the outer and inner walls, the temples and the ziggurats built of brick, and dumped the rubble in the Arahtu canal. And after I had destroyed Babylon, smashed its gods and massacred its population, I tore up its soil and threw it into the Euphrates so that it was carried by the river down to the sea."[35] The fate of Babylon is reminiscent of what was done to Hiroshima and Nagasaki in our time, though thanks to modern technology, with much less investment of time and labor.

From the second millennium B.C. to the second millennium A.D., nomads—whether the Huns who conquered Rome, the Moguls who conquered India, or the Mongols who conquered almost all of Eurasia—were the scourge of all civilizations. As long as the weapons of the barbarians were equal to those of the standing armies defending the tributary civilizations, the superior mobility of the

nomad horsemen enabled them to break through and overrun any
civilization weakened by internal deterioration and discord; and all
civilizations sooner or later experienced the "decline" that inexorably followed the "rise." Strong rising states would push nomad
peoples back onto their marginal steppelands only to find themselves overwhelmed by these same or succeeding nomadic peoples
a century or two later in a period of economic, political, and military
disarray. This nomad-settler, push-pull tension was a central mechanism in several thousand years of human history.

So powerful and horrifying was the imprint these nomad invaders
left on their times that even in far-off St. Albans, near London, the
monk Matthew Paris described the Mongols by reputation in the
following bloodcurdling terms in his chronicle:

> Swarming like locusts over the face of the earth, they have brought
> terrible devastation to the eastern parts of Europe laying it waste with
> fire and carnage. After having passed through the lands of the Sarac-
> ens, they have razed cities, cut down forests, overthrown fortresses,
> pulled up vines, destroyed gardens, killed townspeople and peasants.
> If perchance they have spared any suppliants, they have forced them,
> reduced to the lower condition of slavery, to fight in the foremost
> ranks against their own neighbours. Those who have feigned to fight,
> or have hidden in the hope of escaping, have been followed up by the
> Tartars and butchered. If any fought bravely [for them] and con-
> quered, they have got no thanks for reward; and so they have misused
> their captives as they have their mares. For they are inhuman and
> beastly, rather monsters than men, thirsting for and drinking blood,
> tearing and devouring flesh of dogs and men, dressed in ox-hides,
> armed with plates of iron, short and stout, thickset, strong, invincible,
> indefatigable, their backs unprotected, their breast covered with ar-
> mour; drinking with delight the pure blood of their flocks, with big,
> strong horses, which eat branches and even trees, and which they
> have to mount by the help of three steps on account of the shortness
> of their thighs. They are without human laws, know no comforts, are
> more ferocious than lions or bears, have boats made of ox-hides,
> which ten or twelve of them own in common; they are able to swim
> or manage a boat, so that they can cross the largest and swiftest rivers
> without let or hindrance, drinking turbid or muddy water when blood
> fails them [as a beverage]. They have one-edged swords and daggers,
> are wonderful archers, spare neither age, nor sex, nor condition. They
> know no other language than their own, which no one else knows; for
> until now there has been no access to them, nor did they go forth from
> their own country; so there could be no knowledge of their customs
> or persons through the common intercourse of men. They wander

about with their flocks and their wives, who are taught to fight like men. And so they came with the swiftness of lightning to the confines of Christendom, ravaging and slaughtering, striking everyone with terror and incomparable horror.[36]

The push-pull relationship between nomadism and civilization was only snapped in early modern times when the technological-scientific revolution in the West destroyed the parity in military technology that had existed for so long. How decisive a turning point in human history this proved to be was demonstrated by the Russians, who had lived under Mongol rule for centuries. In the sixteenth century, however, using Western firearms and artillery, the armies of Tsar Ivan the Terrible overthrew the Mongol khans and proceeded to conquer the steppes from the Caspian Sea in the south and to the Ural Mountains in the east. His successors used Western railway and arms technology to push across Siberia to the Pacific, and across the Central Asian deserts to Afghanistan and India.

The outstanding military figures henceforth would not be mounted warriors like Attila and Genghis Khan, but the commanders of the most advanced technologies like General Horatio Kitchener at Omdurman, and General Douglas MacArthur aboard the battleship *Missouri* in Tokyo Bay. Thus, the nomads disappeared from the historical stage that they had dominated for thousands of years, losing their historic role, inevitably and irretrievably, when the horse and the crossbow were superseded by the railway and the machine gun.

CHAPTER 3

CAPITALIST SOCIETIES

It would not be worth my while to make [my steam engines] for three counties only, but I find it well worth my while to make for all the world.

MATTHEW BOULTON (1769)

Always we are hoping that we need expand no farther, yet ever we are finding that to stop expanding would be to fall behind, and even today the successive improvements and inventions follow each other so rapidly that we see just as much yet to be done as ever. When the manufacturer of steel ceases to grow he begins to decay, so we must keep extending.

ANDREW CARNEGIE (1896)

In order to compete effectively in the industry today, you have to be a global player.

ROBERT E. MERCER, chairman of Goodyear (1988)

FROM TRIBUTARY SOCIETY TO
CAPITALIST SOCIETY

The essence of capitalism has been best summarized in Joseph Schumpeter's memorable characterization of its dynamics as "creative destruction."[1] Capitalism's basic competitive drive brings on bursts of technological and institutional innovation, and with these comes increased productivity. So compelling is its creative impulse that capitalism has enveloped the entire globe, overwhelming traditional cultures and economies, while transforming peoples on all continents. This seemingly irresistible process, however, generates the other half of Schumpeter's equation—destruction. Destruction is the inevitable concomitant of relentless creativity. The self-generating technological exuberance and economic expansionism have overrun any institution and any object, animate or inanimate, standing in the way of capitalism's sacred principle: profit or perish. The victims of capitalism's creative drive have included not only overseas food-gathering tribal peoples, but the venerable civilizations of China, India, and the Middle East, and the underdeveloped societies of today's Third World; not only the green valleys of eighteenth-century Wales, but the continental expanses of Amazonia, and the entire planetary ecosystem of the late twentieth century. The particular combination of creativity and destruction that capitalism has generated provides the foundation both for the extraordinary achievements and the appalling setbacks of recent centuries, for the unprecedented promise and peril of our own time.

How successful capitalism's creative destruction has been can best be appreciated in light of the exhaustion and impotence that had overtaken the preceding tributary civilizations. All of them had come to suffer from a common ailment, technological stagnation, depending as they did on the productivity of peasants and artisans, which had remained largely unchanged through the millennia. The British economist John Maynard Keynes perceived the relative immobility of tributary states when he noted that many important inventions "before the dawn of history," had been followed by "remarkable" sterility between "the prehistoric age and comparatively modern times."[2]

Keynes's observation was fully justified. The Neolithic age, the period between the adoption of agriculture and the rise of tributary civilizations, had been technologically precocious. Pioneer villagers had invented the wheeled cart and the plow, discovered the processes of metallurgy and ceramics, and calculated an accurate solar calendar. With the advent of tributary civilizations, this headlong progress slowed drastically. With the exception of the discovery of iron, of coinage and of the alphabet, most inventions henceforth would be for waging war. While the Greeks, for example, invented ratchet-equipped catapults as well as "Greek fire," a petroleum-based incendiary for setting aflame enemy ships and siege machinery, such inventions obviously were not wealth-producing and could not solve the basic economic problems of tributary civilizations.

Tributary societies proved basically inhospitable to the sort of technological innovation that might have led to a breakthrough comparable to the adoption of agriculture. Plenty of hands were available—free, serf, or slave—and it was invariably cheaper to put them to work than to develop new machines. Hero of Alexandria in the first century A.D. used his knowledge of steam power to build a gadget that opened temple doors—a pitiful triviality compared to the earth-shaking engines constructed by James Watt in the fundamentally different milieu of eighteenth-century England.

Tributary societies lacked not only incentives for technological change, but also the necessary desire for such innovation. The elites of those class-based civilizations regarded such matters as beneath their dignity. They studied astronomy and foretold eclipses, yet for centuries failed to invent such simple devices as the wheelbarrow and the horse harness, which would have instantly multiplied the effectiveness of human power and horsepower.

The often-noted cyclical nature of the history of all tributary civilizations can be traced to technological stagnation. Ten peasants were needed simply to support each city dweller, and a large proportion of the rural surplus was squandered by elites on wars, bureaucratic structures, monumental architecture, and opulent living. This is why tributary societies so often were on the brink of bankruptcy. There were a few rare exceptions to this pattern. A burst of new mercantile energy under the Sung dynasty in China (960–1279) led, for instance, to an urban culture and economy that outshone anything Europe had to offer until the nineteenth century. Usually, however, the pressure of burgeoning military and bureaucratic establishments threatened to overwhelm the chronically strained productive capacities of tributary societies. Hence, a vicious cycle of linked events would arise in which rising taxes and an increasing impoverishment would spark popular uprisings culminating in internal revolts, external invasions, or a combination of the two.

The universality of the pattern of imperial rise and fall has led historians to put forward theories about the inevitability of such cyclic movements in history. Comparisons have even been made between the birth, growth, decline, and death of civilizations and the human biological life cycle. But there was nothing either inherent in human nature or biologically inheritable in the actual determining factor that led to the paralysis and decline of one tributary society after another. The limits on technological development fostered by tributary societies simply caused a millennia-long bottleneck. Only with the emergence of capitalism did a new social environment favorable to technological progress release the forces of "creative destruction" that cracked open the tributary world forever.

The central fact of modern world history is that capitalism emerged first in western Europe, which explains that region's rise from obscurity to global hegemony. Western Europe's pioneering role is taken for granted today. Yet how surprising that role actually was, given the fact that that western Europe had been the most backward area in all of Eurasia during the so-called "Dark Ages" following the Fall of Rome. Only now can we see how that retardation later made it possible for Europe to take the lead in social and technological innovation.

The Fall of Rome was not a unique event in the annals of world history. Similar "falls" had occurred repeatedly across the globe.

What made the case of Western Europe unique was that a new imperial "rise" did not follow the "fall." This was in striking contrast to contemporary China where the Han dynasty, too, collapsed, but was followed, with relatively brief interludes, by other imperial dynasties, so that Chinese civilization continued to flourish with little interruption almost to our own time. In Europe, however, repeated attempts at an imperial restoration failed, partly from certain internal weaknesses specific to the area, and partly from the chaos and destruction caused by a ceaseless succession of nomadic invasions. Thus, the Avars and Slavs destroyed the emerging Merovingian Empire in the sixth century, and the Magyars, Vikings, and Muslims had the same effect on the emerging Carolingian Empire in the ninth century.

The repeated demolition of embryonic succession empires in the West cleared the ground for something new to take root. A new Western civilization gradually took shape in which several potentially competing institutions—feudalism in the realm of politics, the papacy in the realm of church-state relations, and manorialism in the realm of economics—replaced the monolithic imperial structure that had been so inhospitable to technological progress. Feudalism meant that a congeries of warring feudal kings and lords replaced the former imperial authority; the papacy meant that an independent church replaced dictation by an emperor; manorialism meant the emergence of autonomous local economic units in place of slave plantations integrated into a vast imperial economy. In addition, a new merchant class slowly emerged in the spaces created by these competing institutions, and gradually gained enough economic and political strength to challenge first the feudal nobles and eventually even the monarchies themselves.

This "pluralism" gave Western society a globally unique openness and dynamism, which in turn generated extraordinary technological precociousness. Frontier conditions and underpopulation stimulated the invention of labor-saving devices and endowed manual labor with a status and respect lacking in the old imperial society. The traditional gulf between artisan and philosopher began to be bridged. Instead of perching on top of pillars, monks in monasteries began to preach that *laborare est orare*—"to work is to pray." Those monks were the first intellectuals to get dirt under their fingernails, and in so doing they contributed to Europe's material progress. Customarily, they were given large tracts of backcountry wasteland, where they set to work to improve their properties. They leveled forests, drained marshes, constructed roads, and built bridges and milldams. Inside their monasteries, the monks practiced skilled crafts, working with leather, textiles, and metals. This economic activity, together with the security afforded by church authority, attracted settlers and artisans, thus making these ecclesiastical centers key incubators of economic and technological progress. The result was that the medieval West made more technological progress than classical Greece and Rome had during their entire histories.

Agricultural productivity increased sharply with the development of a "three-field" rotation system of farming, and with the invention of the heavy-wheeled plow and of an improved horse harness that enabled an animal to pull five times more weight. The watermill and windmill, both known but little used in Greco-Roman times (because of the abundance of cheap labor and the scarcity in the Mediterranean basin of rivers with year-round full water flow), were now developed into power sources for forge hammers, forge bellows, sawmills, fulling mills that produced cloth, pulping mills that produced paper, and stamping mills that produced ore. The five thousand mills listed in England's *Domesday Book* of 1086 represented a substantial increase both in national energy resources and in national goods production, which doubtless raised living standards.

Technological advance stimulated demographic growth. Western Europe's population increased 50 percent between the tenth and fourteenth centuries, a rate that seems insignificant today but that was unmatched at the time. Peasants poured into the vast underpopulated regions of eastern Europe—a medieval European eastward immigration comparable to the one that swept westward into the Americas centuries later. New mining methods raised the salt, silver, lead, zinc, copper, tin, and iron-ore output in central and northern Europe, while the timber and fishing resources of northern Europe began to be exploited more effectively. A general rise in agricultural, mining, fishing, and forestry productivity stimulated a corresponding growth in commerce. Merchants had trafficked in the early Middle Ages, but it had been mostly a small-scale trade in luxuries such as jewelry, silks, spices, and elegant ornaments for churches. By the fourteenth century, a mass trade had developed in necessities such as raw wool from England, woolen cloth from Flanders, iron, timber, and rye from Germany, leather and steel from Spain, and fish from the Baltic.

Such was the general ferment that spawned the new economic order of Commercial Capitalism. Essentially, this involved a novel use of money as capital to make profits and to finance indefinite expansion. No precapitalist society had been based on such a notion of growth. Their aim had always been to restore or maintain rather than to enhance the material standards of the past. With capitalism the goal became precisely the opposite. The "capitalizing" of profits, or the plowing of them back into new enterprises to provide more capital for more production, was the basic rationale behind the term "capitalism." The new "capitalist" was not content to make a living, but was driven by the threat of competitors to maximize his profits and investments in an endless spiral of self-generating growth.

In agriculture such growth was achieved by entrepreneurs who used their capital to displace tenants from their small plots, and to combine the plots into "enclosures" that yielded rich profits through sheep raising or more efficient large-scale farming. In the crafts, the new capitalism meant the substitution of the "putting-out" system for the traditional medieval guilds with their hallowed concepts of "just price" and "just profit." Guild members for centuries had operated on the premise that profiteering at the expense of neighbors was both unethical and unchristian. But here also the new figure of the entrepreneur appeared, using his capital to purchase and distribute (or "put out") raw materials to cottage workers who prepared the finished products, which the entrepreneur bought at minimum rates and sold for maximum profits. In the world of finance, the church's ancient prohibition against charging interest on loans as "usury" and a sin came to be thoroughly ignored by bankers, who observed realistically, "He who takes usury goes to hell; he who does not goes to the poorhouse."

In this way all phases of daily life were monetized according to the ruling principle: profit or perish. Profit could be made by entrepreneurs either by reducing what they paid their workers (on land or in the shop), or by increasing productivity through improved technology. The first strategy, though much used, had its obvious limits since an irreducible minimum level of pay was necessary simply to keep a work force on its feet. By contrast, technology could be improved indefinitely, opening up boundless possibilities for greater productivity and greater profits. This realization led to the unprecedented technological upsurge that accompanied the advent of capitalism and that persists to the present day.

Because of its automatic and unceasing infusion of capital into the economy, Schumpeter concluded that "capitalism not only never is, but never can be stationary."[3] Indeed, history shows that capitalism has expanded its scale of operations inexorably from a local to a national to an international, and now to a planetwide level. Today, corporations are even attempting to extend their operations into space.

Capitalist dynamism has expressed itself not only geographically but also institutionally, as capitalism has undergone successive transformations. It is possible to identify three transformational stages in its history: Commercial Capitalism, 1500–1770; Industrial Capitalism, 1770–1940; and High-Tech Capitalism since 1940. With each of these phases, capitalist creativity and destructiveness have intensified.

COMMERCIAL CAPITALISM, 1500–1770

CREATIVITY

In the fifteenth century, the western Europeans launched their great voyages of discovery, which for the first time brought the peoples of all continents into direct contact with each other, thereby beginning the global phase of human history. Why was it that western Europeans pioneered in this fateful enterprise? Why didn't the Chinese "discover" Europe, since medieval Chinese technology had been far ahead of Europe's, as had its navigational skills. This is no trivial question. If a Columbus and a da Gama had set forth from China, and if Chinese had colonized the Americas, Australia, and the rest of Oceania, the population of the globe today might be at least one-third rather than one-fifth Chinese.

One reason why Columbus and da Gama were not Chinese is evident from contemporary European and Chinese writings. "Coming into contact with barbarian peoples," wrote the historian, geographer, and mariner Chang Hsieh in 1618, "you have nothing more to fear than touching the left horn of a snail. The only things one should really be anxious about are the means of mastery of the waves of the sea—and, worst of all dangers, the minds of those avid for profit and greedy of gain."[4] By contrast, when Columbus landed in the Bahamas, he reported that the native Arawak were gentle and loving and "remained so much our friends that it was a marvel." But this same Columbus soon was writing back to Spain: "From here, in the name of the Blessed Trinity, we can send all the slaves that can be sold. . . . for these people are totally unskilled in arms." Columbus further promised that he could supply his sovereigns with "as much gold as they have need of, and in addition spices, cotton, mastic and aloes-wood." A few decades later, conquistador Hernando Cortés, whose small force of conquistadors overran the Aztec Empire, wrote candidly about his motives in going to the New World: "We, the Spanish, suffer an affliction of the heart which can only be cured by gold. . . . I came in search of gold and not to work the land like a laborer."[5]

How different were the motives that drove the Europeans from those of the Chinese. While a profit-or-perish ethos impelled Europeans to reach out over oceans and to penetrate continents, the Chinese were warning that greed for profit was the "worst of all dangers." This fundamental difference between capitalist Europe and the rest of the world explains in large part not only why the Europeans reached the Americas and the East Indies, but also why their descendants persevered in their expansionism until they dominated the entire world, including ancient Asian empires that were more populous and wealthy than western Europe's states. This is why some historians now hold that the most important figures in Europe's overseas expansion were not famous explorers like Columbus and da Gama, but entrepreneurs with their capital resources. Although they never ventured out of their home ports, it was they who were "responsible for the foundation of many of the colonies . . . who kept the colonies supplied . . . who opened new markets, sought new lands, and enriched all of Europe."[6]

Such was the creative force of Commercial Capitalism, so-called because between the sixteenth and eighteenth centuries most of the capital was invested in commerce. These were the centuries when western Europe emerged from its medieval inferiority and obscurity to establish its superiority in field after field of human endeavor.

The New Thinking

A British consul and merchant who lived in the Ottoman Empire in the late eighteenth century noted with scorn the superstition and sterility that anesthetized the Ottoman arts and sciences. He reported that "from the mufti to the peasant" it was believed that the earth was "suspended by a large chain," that the sun was a ball of fire the size of an Ottoman province, and that the eclipses of the moon were caused by "a great dragon attempting to devour the luminary."[7] While the consul's disdain was understandable, had he been at the University of Paris a few centuries earlier, he would have found the professors there disputing about the number of teeth in a horse's mouth. All they could agree on was that the number could not be a multiple of three, for that would be an offense to the Trinity; nor a multiple of seven, for God created the world in six days and rested on the seventh. Neither the records of Aristotle nor the arguments of St. Thomas Aquinas enabled the Paris savants to solve the problem.

The ignorance and unreason rampant in the eighteenth-century Ottoman Empire struck the British merchant as contemptible only because in the preceding centuries the Scientific Revolution had radically changed ways of thinking in the West, and in the West alone. "Philosophers" there had begun to consider it part of their work to look in horse's mouths and actually count the teeth there (or at least to consort with those of humbler birth who had done so). The nature of that intellectual transformation is reflected in the charter King Charles II of England granted in 1662 for the establishment of "The Royal Society of London for Promoting Natural Knowledge." The objective of the society was to seek knowledge "not only by the Hands of learned and professed Philosophers; but from the Shops of Mechanicks; from the Voyages of merchants; from the Ploughs of Husbandmen."[8]

This objective of bringing together "Philosophers" and "Mechanicks" was the secret behind the intellectual triumph of the West. During millennia of tributary civilizations, the gulf between high and low cultures had effectively separated artisans from scholars, so that settling an argument by simply going to a stable and starting to count a horse's teeth became inconceivable. Upper-class intellectuals, or "philosophers" as they were called, considered it demeaning to indulge in anything remotely resembling manual labor. Their mission, as they saw it, was to think with their minds, not to soil their hands.

Such compartmentalization of thought and activity lost its power and purpose given the economic needs of commercial capitalism. Oceanic expeditions created a demand for larger, sturdier, more maneuverable ships, which was met by combining shipbuilding techniques originally developed in the Middle East, the Mediterranean, and northern Europe. Between 1200 and 1500 the tonnage of the average European ship doubled or trebled, and it became capable of riding out oceanic storms that would have sunk a medieval vessel. Armed with new naval artillery superior to any in the world, European merchant ships and warships quickly won mastery of the oceans and of global commerce.

At the same time, comparable advances were being made in the art of navigation. So long as ships remained in the Mediterranean or sailed only north-south along the coasts of Europe and Africa, latitude could be determined by age-old astronomical methods and also by the simple and cheap quadrant, which was widely used by the sixteenth century. But when mariners began to cross oceans east to west, an accurate determination of longitude became essential. Rough estimates were made by means of the hourglass, but precise reckoning had to await Galileo's discovery of the principles of the pendulum in the seventeenth century. Navigators also were aided by advances in the art of mapmaking and by up-to-date compilations of nautical information. The latter were refreshingly objective and matter-of-fact compared to earlier writings in which learned doctrinaires speculated whether humans could survive killing sunbeams and boiling ocean water at the equator.

Meanwhile, on land, expanding mining operations stimulated advances in pumps and power transmission as well as in mechanical and hydraulic principles. Mining also brought to light new ores and new metals like bismuth, zinc, and cobalt. Techniques for separating and handling these had to be found by protracted trial and error. In the process, a general theory of chemistry began to take form, involving oxidations and reductions, distillations and amalgamations.

The knowledge gained in these and other fields was widely disseminated by the printing press, which made available ever-cheaper books to an ever-growing audience. Ambitious craftsmen were able to study in books the accumulated experiences of others and to apply them to their own problems.

The significance of this generalized interaction and cross-fertilization was recognized at the time, especially by the great English exponent of science Francis Bacon. In his famous *Novum Organum* (1620), he called on his readers to "renounce notions, and begin to form an acquaintance with things," turn to mechanics for fresh knowledge, check theories with experiments, and reappraise accepted tradition, especially from ancients such as the Greeks, whom he excoriated for not having conducted a single experiment. Especially revealing is Bacon's imagery of himself as the Columbus of a new intellectual world, sailing through the Pillars of Hercules (a symbol of the old knowledge) into the Atlantic Ocean in search of new and more useful knowledge. In fact, he explicitly stated that "by the distant voyages and travels which have become frequent in our times, many things in nature have been laid open and discovered which may let in new light upon philosophy."[9]

Just as Bacon grasped the interaction of commercial enterprise and "philosophy," so another English author, analyzing the Industrial Revolution of his time in the early nineteenth century, sensed the interaction of science and technology and the resulting mutual stimulation: "The manufactory, the laboratory, and the study of the natural philosopher, are in close practical conjunction. Without the aid of science, the arts would be contemptible: without practical application, science would consist only of barren theories, which men would have no motive to pursue."[10]

It is no exaggeration to conclude that the new way of thinking embodied in the Scientific Revolution, which began in the sixteenth century and persists to the present, altered fundamentally the intellectual balance between the West and the rest of the world. In the past, Westerners had been dismissed by the Chinese as "long-nosed barbarians," and by the Muslims as ignorant infidels. Following the Scientific Revolution, beseiged and beleaguered ruling elites all over the world began to search for the "secret" of Western technological superiority as they tried to cope with Western economic and military aggression backed by unheard-of machines of all sorts.

How great this reversal in attitude was can be seen in the words of a Muslim Uzbek leader who felt constrained to tell his followers that science "is the cause of the progress of a people," that it "led the savage Americans to their present high status," that it reveals "the secret meaning of the Koran," and that it could "free our nation from the yoke of the infidel and restore Islam to its earlier heights."[11]

The New Politics
New thinking under Commercial Capitalism was matched by a new politics as Europe's monarchs built cohesive and aggressive state structures. These monarchs were aided by a growing merchant class that provided financial support for, and competent officials to staff, state bureaucracies. In return, a newly consolidated royal power offered these merchants an atmosphere conducive to commercial expansion by ending the incessant feudal wars that had ravaged Europe, and by enforcing uniform national laws, weights, and currencies. By the end of the Thirty Years' War (1618–48), the individual sovereign state was recognized as the basic unit in European politics.

Just as the global intellectual balance shifted because of the Scientific Revolution, so the global political balance shifted because of the new aggressiveness of the European states. So many of them existed within the restricted confines of western Europe that they were intensely competitive, fighting each other for land within Europe and also overseas. The contrast between European expansionism and Chinese withdrawal from overseas operations is striking and instructive. Between 1405 and 1433, China's Ming dynasty launched a series of remarkable maritime expeditions that rounded Southeast Asia and pressed on as far as Africa and the Red Sea. These voyages were organized and led by a court eunuch, whose ships returned laden with zebras, ostriches, giraffes, and other exotic animals for the titillation of the imperial court. When that court lost interest, the expeditions were ended by imperial order, a denouement that would have been utterly inconceivable in contemporary Europe with its rival national monarchs and merchant companies competing ferociously in their respective overseas enterprises. Indeed, when the Spanish and Portuguese courts laid claim to divide and possess all overseas territories, France's King Francis

I retorted: "I should like to see Adam's will, wherein he divided the earth between Spain and Portugal."

The aggressiveness of the European states was enhanced by their superior cohesion, which gave them a decisive advantage over tributary civilizations. While there was discord within all of the European states, the fatal chasm between top and bottom, between rulers and ruled, that left Asian and Amerindian empires so disastrously vulnerable, was nowhere apparent. One reason why a few hundred Spanish adventurers conquered the great Aztec and Inca empires with such ease was that thousands of Indians previously subjected and oppressed by those empires fought beside the conquistadors. In India a handful of British merchants won control of a subcontinent many times the size and population of their own country, primarily because India was not an integrated state but a disparate agglomeration of varied ethnic, religious, and social elements that could be turned against one another. When the Indian Mutiny broke out against British rule in 1857, it was suppressed by Indian as well as British troops. Just as British rule in India owed much to native sepoy mercenaries, so French rule in North Africa rested heavily on native Zouaves, predecessors of the Foreign Legion, who were organized in 1831 and drew mercenary recruits from all over the globe. Not only did the British and French exploit local rifts to recruit mercenaries, but they also deliberately aided and abetted those rifts in pursuit of successful divide-and-rule strategies. In India the British played off Hindus against Muslims; in North Africa the French pitted Arabs against Berbers; and throughout sub-Saharan Africa all the European colonialists favored selected districts or ethnic groups with special privileges denied to others.

The New Economics

Alongside the new politics, a new economics emerged in Europe, the two being interactive as well as concurrent. In an unparalleled burst of innovation and expansionism, this new economics dismantled and restructured institutions and ways of life, both within Europe and overseas.

Agriculture was monetized, so that land now came to be viewed as a potentially profitable business investment rather than as a means of support for a local population. The resulting "enclosures" of small plots traditionally held by peasants generated a greater output of foodstuffs and also released manpower that was needed in the growing cities. Craft guilds were short-circuited by new entrepreneurs who boosted productivity through the use of cottage industry, where the guiding principle was "maximum profit" rather than "just price." In commerce, the old merchant guilds were overshadowed by newly formed joint-stock companies.

These companies efficiently mobilized relatively enormous sums of capital for national and international operations because they enabled anyone to speculate while limiting their liability—their possibility of loss—only to what they individually invested in company shares. Details of management were entrusted to directors selected for their experience, and these directors in turn chose dependable individuals to manage the company's specific overseas projects. The joint-stock institution made it possible for European capital to penetrate the entire globe under the aegis of the Dutch, English, and French East India companies, the various Levant and Africa companies, the Muscovy company, and the still extant Hudson's Bay Company. No Eastern merchant, limited to his own resources or those of his partners, and choosing his managers from his family or circle of acquaintances, could hope to compete with such powerful and impersonal organizations.

The resultant boom in Europe's global commerce was accompanied by a body of economic theories and practices known as mercantilism. The objective of mercantilism was to strengthen the European monarchies by amassing bullion—gold or silver, uncoined or in mass, as in bars or plates—through a favorable balance of trade, by seizing or developing colonies that first could be looted of their wealth for the king's treasury and then would produce raw materials needed by the mother country. It was for these reasons that in country after country the monarchy granted royal charters to joint-stock companies for trading in or colonizing specified overseas territories. This, in turn, led to a worldwide competition to establish trading posts and colonial settlements. The prime objective of company officials was to produce a maximum profit for their shareholders, so after initially pillaging the most obvious sources of local wealth, whether carrying off Amerindian bullion or taxing Bengali, company officials soon sought to develop their new colonies as cornucopias of raw materials—sugar in the West Indies, tobacco and cotton in the American South, tea and jute in India, and coffee and rubber in Southeast Asia. The production of such commodities required large labor forces—first obtained from local people, later from European indentured servants, and finally from African slaves.

If the Chinese with their enormous population had desired to make the colonization of Southeast Asia official policy after the fashion of the western Europeans, they could have established settlements with populations much greater than the 2 million in England's thirteen colonies in North America. But the noncapitalist nature of China's society meant that it never could have even occurred to its leaders to wage wars for overseas possessions, or to promote and finance the emigration of its citizens. How incredulous Chinese officials would have been had they been informed that in 1739 England and Spain fought the War of Jenkins' Ear, precipitated by the determination of British merchants to trade freely with Spanish colonies in the Caribbean. One of these merchants, Captain Robert Jenkins, was attacked near Jamaica, his ship seized, and one of his ears severed in the ensuing melee. The Chinese would also have been incredulous to learn that Jenkins wrapped his ear in cotton and displayed it throughout England and before the House of Commons in a successful campaign to declare war on Spain and force open the Spanish West Indies to British trade. Bells throughout the country rang out in celebration when King George II formally proclaimed in 1739 that hostilities with Spain had begun.

The same king displayed similar commercial zeal when he sent a trade mission to the court of Emperor Chien Lung to arrange for more trade between their two countries. The fundamental societal difference between China and Britain is evident in the emperor's condescending response to the king: "There is nothing we lack, as your principal envoy and others have themselves observed. We have never set much store on strange or ingenious objects, nor do we need any more of your country's manufactures. . . . This [trade mission] is indeed a useless undertaking."[12]

Despite temporary rebuffs, the competing European states succeeded in peppering the coasts of the New World, Africa, and Asia with their colonial settlements and trading posts. That this process of global colonization happened as it did is now so taken for granted that we can imagine no other course to world history. We tend to forget how specifically the process was driven by the needs of Commercial Capitalism. Yet during the same centuries that Europeans were planting their settlements from Newfoundland to the tip of Patagonia, southern Chinese were beginning to emigrate in large numbers to Southeast Asia. How different the result. China's imperial government not only did nothing to encourage the settlers but actively discouraged them. Emigrants were specifically forbidden to leave their homeland, and when they were persecuted, even massacred in their new homes, the imperial government pointedly ignored their plight. How different it was in the New World where by the time of the American Revolution the population of the English colonies amounted to 2 million, roughly a fifth of the total population of the English-speaking world.

The new economics of Commercial Capitalism led to the emergence of an integrated, more productive global economy. Before 1500, Arab and Italian merchants had carried on a long-distance trade across the expanse of Eurasia, but on a relatively small scale, and mainly in luxuries such as spices, silks, perfumes, and precious stones. By the late eighteenth century, this limited Eurasian commerce had been overshadowed by a large-scale, triangular trans-Atlantic trade in which rum, guns, cloth, and other manufactured goods went from Europe to Africa to pay for slaves to work the New World plantations that produced sugar, tobacco, and other agricultural goods for Europe.

Almost as important to the new global trade were the grain, minerals, cattle, and lumber being shipped from eastern to western Europe, and the textiles and other manufactured goods moving in the opposite direction. At the other end of the world, silver mined in Spanish America was being shipped directly across the Pacific via galleon to Manila, and thence to China to pay for silks and porcelains. Between a third and a half of the entire amount of bullion produced by American mines is believed to have ended up in the Far East. During the seventeenth century, as many as forty-one Chinese ships called each year at Manila, exchanging the prized products of Chinese crafts for American silver. In this manner, a truly global economy was emerging for the first time.[13]

These new worldwide trade patterns resulted in a corresponding global diffusion of plants and animals. Eurasian grains (wheat, rye, oats, and barley) and animals (horses, cattle, and sheep) were transported to all continents, and a remarkable store of domesticated Amerindian plants (potatoes, corn, peanuts, and several varieties of beans, pumpkins, and squashes) was carried around the world. The distribution of these plants and animals across the globe meant a substantial increase in the world's food supply, which in turn led to a corresponding increase in world population. China's population, for example, seems to have remained fairly steady for several centuries, but increased between 1540 and 1640 from about 60 million to 150 million, thanks largely to the introduction of such New World plants as corn, the peanut, and the sweet potato. World population is estimated to have risen from 427 million in 1500 to 641 million in 1700 to 890 million in 1800.

Not only did the new global trade pattern help increase the supply of necessities, it also increased the availability of new amenities. This was especially true in western Europe, which initiated, directed, and profited the most from the new global economy. Coffee, tea, and cocoa became common beverages and, together with tobacco, common lubricants of social life.

DESTRUCTION

The intellectual, political, and economic creativity of Commercial Capitalism was matched by a destructiveness especially obvious in those non-European lands where local peoples frequently found themselves relatively defenseless before an enveloping capitalist tidal wave. The exact degree of this vulnerability and destruction depended on local and regional levels of technological development and social cohesion, which determined the effectiveness of military resistance against European arms, and on the degree of geographic isolation from other human communities, which determined the effectiveness of biological resistance against European diseases. Most vulnerable, consequently, were Australia's aborigines, who were the most isolated geographically and the least developed technologically. After them came the Amerindians of the New World, the Africans, and finally the Asians, who were both technologically advanced, geographically the most centrally located, and therefore the most impervious to European arms and diseases.

Because of the extreme isolation of Australia in the southern Pacific, the Europeans did not reach that continent until the late eighteenth century. The Amerindians, by contrast, felt the full blast of capitalist destructiveness from the moment Columbus made his fateful landing in the West Indies. The significance of capitalism as a factor becomes apparent if it is recalled that the Vikings had stumbled on North America approximately five hundred years before Columbus, and had for a century tried unsuccessfully to maintain settlements there. By contrast, what followed Columbus was a massive and overwhelming penetration of both North and South America by peoples from every country in Europe. The difference in impact between the two periods is to be found in the economic development of Europe during the intervening half millennium, and in its new spirit of expansionism and acquisitiveness. "One who has gold," observed Columbus, "does as he wills in the world, and it even sends souls to Paradise."[14]

Whatever may have befallen Columbus in paradise, he, as well as all later western Europeans more or less did as they willed in the New World. Despite the size, wealth, and impressive attainments of the Aztec and Incan empires, they were easily toppled by a few hundred Spanish adventurers. The most basic reason was that the Amerindians had been isolated from the rest of humankind since they crossed the Bering Strait more than forty thousand years earlier. This isolation left them relatively defenseless against Europe's new weaponry and totally defenseless against the smallpox, measles, and typhoid they brought with them, as well as the malaria and yellow fever of their African slaves. As a result, the population of Spanish America dropped from an estimated 50 million at the time of the conquest to 4 million by the seventeenth century.

Alcohol and brutal exploitation in mines, on haciendas, and on plantations further decimated the Indian survivors of this biological holocaust. So overwhelming was the destruction the European intrusion brought about that the New World today is peopled by a majority of whites, with minorities of blacks, Indians, mestizos (European-Indian mix) and mulattoes (European-African mix) in that order. In addition to this racial displacement, both North and South America were completely partitioned among the European states. Spain acquired all of Central and South America, except for Portuguese Brazil. North of the Rio Grande, Britain planted her Thirteen Colonies, which eventually became the United States, and north of the Great Lakes was the colony of New France, which today is part of Canada. So far as the original Amerindian inhabitants were concerned, their historic fate was to endure the "destruction" inherent in expansionist Commercial Capitalism—a destruction that Adam Smith summarized at the time in the phrase "dreadful misfortunes."

The fate of the Africans was somewhat less immediately cataclysmic than that of the Amerindians. A prime reason was that they had not been so completely isolated from Eurasia during the preceding millennia, and therefore had developed immunities to the diseases that decimated the Amerindians. In fact, until nineteenth-century advances in tropical medicine offered some protection, it was the Europeans arriving on the west coast of Africa who were decimated by African diseases as the Amerindians had been by those of the Europeans. Past contacts with Europeans had also given Africans knowledge of the firearms that on first contact had so overawed the Amerindians.

How much less vulnerable the Africans were became clear as the Europeans sought the chief commodity they desired from Africa— slaves. This European drive for slaves was comparable commercially to their drive for furs in North America, but while the Indians could not stop the Europeans from spreading across their continent, monopolizing all stages of the fur trade, the Africans insisted on managing the slave trade themselves. Thus, they won the rich profits of middlemen by collecting slaves and selling them to European slave traders at coastal ports in return for rum and manufactured goods, including guns.

Despite this modicum of autonomy, the African continent as a whole was ravaged by the slave trade. The precise impact varied from region to region. The coastal Dahomey and Ashanti, who raided or purchased their slaves from the interior, profited at the expense of their neighbors. By contrast, the Kongo and Angola regions suffered severely from the depredations of the Portuguese who forced their way in and dispensed with African middlemen. Arab slave traders did the same in East Africa, where they rounded up slaves for labor on sugar, rice, and spice plantations on the islands of Madagascar, Réunion, Mauritius, Seychelles, and Zanzibar— commonly referred to at the time as the "West Indies of the Pacific."

The overall pattern was one of economic and social disintegration as warring groups, armed by European traders, engaged in self-perpetuating raids and abductions.

Some African chiefs tried to stop the nefarious selling of fellow Africans to Europeans, but these mavericks were hopelessly vulnerable because if they did not trade for European firearms, they found themselves defenseless against rivals who did. Concerning the average African, a Dutch observer reported in 1816 that the ever-present danger of "any day being seized and sold to the ships extinguished the courage for all peaceful labour, and made the negro into an armed and restless robber who laid snares for his fellowman to catch and sell him, as he feared and expected for himself. . . . One even hears of murders and housebreakings . . . continual quarrels with neighbors."[15] Not only did anarchy breed anarchy, but the rich and easy profits of the slave trade discouraged Africans from engaging in normal economic activities. "This [slave] trade," reported a British observer in 1817, "is beyond all comparison so indolent and lucrative that even were there any appeal to their feelings, it would not influence in competition with such inordinate gain. Every other trade requires, comparatively, activity and exertion, and yields very inferior profit. It is unreasonable, therefore, to expect any conduct on the part of the natives but such as may be auxiliary to the slave traders."[16]

An estimated total of 40 million Africans were captured in the interior of their continent, of which about 10 million eventually ended up working on New World plantations. The other 30 million became casualties of the grueling overland marches from the interior to the coast, or of the dreaded transatlantic crossing. Horrific as it was, Commercial Capitalism's disruption was not as total in Africa as in the Americas, partially because the original African population was larger and partially because the slaves were taken over a period stretching from 1450 to 1870. Consequently, the African continent remains racially African in contrast to the Americas, which are no longer exclusively Mongoloid. But this fact must be balanced by another: about one-third of all people of African descent today live outside Africa, and these are not the descendants of voluntary emigrants, comparable to those who converged on New York's Ellis Island at a later period.

In contrast to Africa and the Americas, Europe's disintegrative impact on Asia took far longer to make itself felt. During the era of Commercial Capitalism, in fact, it was slight simply because Europe lacked the military and economic strength to challenge the still powerful Chinese, Indian, and Ottoman empires. What Europe did enjoy, however, was superior naval power, which allowed the British, French, and Dutch East India Companies to monopolize the lucrative trade between Asia and Europe. The total amount of this trade, however, was much less than the mass transatlantic trade. Consequently, the continent of Asia was relatively unaffected by European capitalism during these centuries. With the sole exception of the spice islands of the East Indies, overrun by the Dutch, the Europeans were easily kept at arm's length. Nowhere on the Asian mainland were they able to make the territorial conquests they had in the Americas, or traffic in human beings as they had in Africa.

In this same period, the destructive power of Commercial Capitalism was manifesting itself far closer to home. The "dreadful misfortunes" Adam Smith noted in overseas lands were also becoming part of the daily lives of European peasants and artisans as they were being separated from their sources of livelihood by entrepreneurs using newly acquired capital to finance land enclosures and cottage industries. As peasants were separated from their ancestral plots and artisans from their protective guilds, the "creative" outcome was increased productivity in the countryside and in urban workshops. But there was also a destructive by-product: displaced peasants and artisans, cut off from their sources of livelihood, left penniless and rootless.

This was the dark underside of capitalism in Europe, which experienced a growing phenomenon of able-bodied vagabonds drifting about, resorting to any measures to stay alive. All past ages and societies had been accustomed to poverty caused by old age, or the sickness or death of breadwinners. In addition, peasants in tributary societies suffered the privation and misery of being forced to yield most of what they produced to ruling elites, with the resulting desperate insurrections and ruthless repressions that punctuated human history. Now, however, western European countries undergoing the transition to capitalism faced a new problem of peasants deprived not of excessive portions of the produce of their plots, but of the ability to work the land itself. The millennia-old burden of extortionate tribute collecting had been replaced by the new and more traumatic shock of outright displacement and uprooting. The plight of the new class of homeless and wandering victims was expressed in the nursery rhyme:

> Hark! Hark! the dogs do bark;
> the beggars are coming to town.
> Some give them white bread,
> and some give them brown.
> And some give them a good horsewhip,
> and sent them out of town.

Some of those who were horsewhipped tried to escape their desperate situation through indentured servitude in New World colonies. In return for transportation to a specified colony, indentured servants promised to work for a master for a fixed period of time, after which they were free to do as they wished. How harsh these arrangements were is evident in the fact that of every ten indentured servants in British North America between 1607 and 1776, only two became independent farmers or artisans after their years of servitude. Most died before their contracts expired; the rest became day laborers or paupers in the New World as they had been in the Old. Equally dismal were the fates of those who were shanghaied to serve as seamen on slave ships plying the Atlantic. Because the seamen had no immunity against tropical diseases, one in five normally died on each crossing, a mortality rate higher than that of the slaves below deck.

The plight of the dispossessed provoked class hatred against the few engaged in the "money business." "God will come to you to punish you for your oppressions," warned a tract of the Levellers, a radical element in the mid-seventeenth-century English Revolution. "You live from the work of other men, but you give them only bran to eat, extorting enormous rents and taxes from your brothers. But what will you do from here on? For the people will submit no longer to your slavery, as the understanding of the Lord enlightens them."[17]

The threat of the Levellers was by no means idle bombast. From the thirteenth century onward, urban revolts and peasant uprisings erupted throughout western Europe, reflecting the widespread dispossessions in rural areas and unemployment in cities. It was in this period that peasants revolted in Flanders and were supported by workers in Ypres and Bruges; that textile workers in Florence seized power; that English peasants under Wat Tyler captured London; and that uprisings broke out in Catalonia, Bohemia, Sweden, Norway, Denmark, and Finland.

Nothing like this was happening in other regions of Eurasia. Contemporary travelers reported that peasants in the Ottoman Empire were better off and more contented than their counterparts across the border in Christian Europe. The explanation was that the Ottoman lands were not experiencing the painful convulsions of capitalism. Yet a time when this equation would be reversed was fast approaching. Already French, English, and Dutch Levant companies were taking control of the trade of the Middle East from local merchants and beginning their merciless exploitation of the Ottoman Empire. Indeed, by 1788, the French ambassador in Constantinople was able to boast with much justification that "the Ottoman Empire is one of the richest colonies of France."[18] In this fashion, Commercial Capitalism manifested on all continents, though to varying degrees, its creative and its destructive impulses.

INDUSTRIAL CAPITALISM, 1770–1940

The transition from Commercial to Industrial Capitalism exemplifies another of Joseph Schumpeter's maxims—that a "stationary" capitalism is an impossibility. Global productivity and trade had increased sharply during the centuries of Commercial Capitalism. England's exports and imports had each risen 500 to 600 percent in the period from 1698 to 1775. This upsurge both reflected and encouraged an increasing global demand for more English manufactures such as textiles, firearms, hardware, ships, and ship accessories including rope, sails, pulleys, and nautical instruments. This growing demand, in turn, forced English industries to improve their organization and technology in a dynamic spiral of demand, invention, and productivity.

Commercial Capitalism provided not only an incentive for technological innovation but the necessary capital to make it possible. Profits poured into Europe from the fur trade in Siberia and North America, the silver mines of Mexico and Peru, the African slave trade, the plantations of the New World, and the East India, West India, Levant, and Africa companies. This combination of incentive and plentiful capital triggered a speedup in the tempo of mechanical invention and its industrial application that came to be known as the Industrial Revolution. Under way first in England by about 1770, and soon after on the Continent, it was responsible for the transition from Commercial to Industrial Capitalism, meaning simply that the bulk of capital began to be invested in industrial rather than commercial enterprises. Where industry had once been an accessory to commerce, the relationship was now reversed. Industrial Capitalism, however, retained a fundamental similarity to Commercial Capitalism in the way its impact combined vast bursts of creative energy with a powerful destructive impulse. Both this creativity and destructiveness were, however, far more global in scope and intense in impact, reflecting the increasing power and dynamism of evolving capitalism, a power to unravel whole societies that persists to the present moment.

CREATIVITY

The New Thinking

New thinking under Commercial Capitalism had taken the form of a Scientific Revolution that owed much to the unprecedented partnership of mechanics and philosophers, of technology and science. At the outset, science was the junior partner, receiving far more from the workshop and the mine than it gave in return. This was equally true in the early stages of the Industrial Revolution, when pioneer inventors almost invariably were talented mechanics. By the mid nineteenth century, however, science began to take the lead in transforming old industries or even creating entirely new ones.

This leading role for science was especially evident in chemistry and later in biology. The origins of chemistry date back to prehistoric times—to the arts of cooking, and of identifying medicinal plants in order to extract drugs from them. Initially in Europe, chemistry's goal was to discover a means to transmute cheap metals into gold, or an *elixir vitae* that would cure all human ills. Antoine Lavoisier (1743–94), who fell victim to the guillotine during the

112

French Revolution, brought order to all the previously chaotic phenomena of chemistry with his law of the conservation of matter: "Although matter may alter its state in a series of chemical actions, it does not change in amount; the quantity of matter is the same at the end as at the beginning of every operation, and can be traced by its weight."

With this formulation, Lavoisier placed chemistry on what we would now consider a solid scientific footing, as Newton had done earlier for physics with his law of gravitation. Lavoisier's successors made discovery after discovery, many with important industrial applications: Justus von Liebig for chemical fertilizers, W. H. Perkin for synthetic dyes, and Louis Pasteur with his germ theory of disease, which led to the adoption of sanitary precautions and so brought under control old scourges like typhoid, cholera, and malaria. These medical advances had profound repercussions, facilitating a rapid population increase, first in Europe and then throughout the world.

In the field of biology, Charles Darwin had an impact that transcended both science and industry. With his doctrine of evolution, Darwin held that animal and vegetable species in their present diverse forms are not the fixed and unchangeable results of separate special acts of creation. Rather they had evolved from a common original source via "natural selection" in a "recurring struggle for existence." The details of Darwin's doctrine have been modified by later research, but virtually all scientists now accept its essentials. Nevertheless, Darwin was widely attacked, and naturally so, for just as the Copernican system of astronomy had deposed the earth from its central place in the universe, so Darwinism seemed to dethrone humans from their central place in the history of the earth.

Despite the critical reception, a much-reduced and simplified version of Darwinism profoundly affected Western society, because the idea of "the survival of the fittest" seemed to confirm so exactly the temper (and power relations) of the times. In politics this was the period when Bismark unified Germany by "blood and iron." In economics this was the period of free enterprise and rugged individualism. In international affairs, this was the age of colonial expansion, worldwide empires, and the "inferiority" of conquered races. In each of these areas, Darwin's theories—or at least popularized versions of them—seemed to provide persuasive rationalizations for the materialism, realpolitik, and racism that were already in vogue. "Social Darwinism," as it came to be called, had in the end little to do with Darwin himself, or with his scientific writings. Yet it was generally accepted because it provided an ideological under-pinning for the new economics and new politics that were enveloping the globe. Social Darwinism was the ideological counterpart of Kipling's dictum:

> That they should take who have the power
> And they should keep who can.

The New Economics

Once the Industrial Revolution got under way, it triggered a chain reaction of continual technological innovation as invention in one industry necessitated invention in others. The pioneer inventions in the cotton industry (John Kay's flying shuttle, Richard Arkwright's water frame, James Hargreaves's spinning jenny, Samuel Cromp-ton's spinning mule, and Edmund Cartwright's power loom) created a demand for plentiful and reliable power to which James Watt's steam engine was a response. The new cotton machines and steam engines required a greater supply of iron, which led to inventions like Abraham Darby's substitution of cheap coke for expensive charcoal in smelting iron ore, and Henry Colt's puddling process, which removed impurities from brittle smelted iron or "pig iron," transforming it into stronger wrought iron preferable for machine manufacture. The need to transport large quantities of ore, coke, iron, and machinery in turn necessitated improved transportation facilities, stimulating successive booms in the building of canals, hard-surface roads, railroad lines that spanned continents, and steamships that spanned oceans.

Economists identify three "long waves" in the evolution of Indus-trial Capitalism. The first in the late eighteenth and early nineteenth centuries centered in the textile, iron, coal, and steam engine indus-tries; the second in the mid nineteenth century was based on rail-ways, steamships, and the mass production of steel; the third in the late nineteenth and early twentieth centuries had at its core the automobile, electrical, communications, and petroleum industries.

The repercussions of these successive economic bursts of activity were profound. First and foremost, Industrial Capitalism evolved self-generating economic growth by systematically harnessing science for specific industrial goals. Laboratory complexes equipped with elaborate and expensive equipment and staffed by trained scientists became integral components of all large industries. Whereas inventions had previously resulted from individual responses to opportunity, now they were planned and virtually made to order. "In modern times," observed Walter Lippmann, "men have invented a method of inventing, they have discovered a method of discovery. . . . We know, as no other people ever knew before, that we shall make more and more perfect machines."[19]

The economic dynamism of modern Europe was not merely the result of a series of technological chain reactions. The "military-industrial complex" (against which President Eisenhower was to issue his famous warning in 1961) made its first appearance in the late nineteenth century. This was a period when military technology was evolving rapidly, a prime example being the launching of a new super battleship, the HMS *Dreadnought,* in Britain in 1906. Each new weapons program stimulated further technological innovation, making older weapons obsolete, and necessitating still larger government appropriations for a new round of arms competition.

The end result was something quite new in history—what William H. McNeill has termed "command technology."[20] Government officials now began to specify the desirable performance characteristics of a new gun, engine, or ship, before it was invented, instead of waiting for random inventions to be submitted by individual technicians or firms. This process was speeded up with the appearance of private firms that pioneered in one new technology after another. This inceptive "military-industrial complex" was to attain its full flowering during and after World War II when the goal of command technology escalated exponentially from a super battleship to the bomb that razed Hiroshima and inaugurated our atomic age.

Self-generating, nonstop economic growth altered fundamentally the economic balance between the newly industrialized states and the rest of the world. For centuries India had led globally both in quality and quantity of textile production. Yet an eighteenth-century Indian hand spinner still required over 50,000 hours to process 100 pounds of cotton. Thanks to Crompton's and Arkwright's inventions, English workers were by 1825 processing 100 pounds of cotton in 135 hours, and the quality of their yarn was superior to equivalent Indian yarn. Equally significant was the simultaneous development in the American South of Eli Whitney's gin, which enabled workers to separate 50 pounds of sticky cotton fibers from seeds in the time previously needed for a single pound—thus assuring a plentiful supply of cheap raw cotton for Britain's mills. It is interesting to note that this crucial machine had first been invented before Whitney's birth by a priest in Mexico, where it was ignored and forgotten, just as centuries earlier the steam engine had been invented and subsequently forgotten in Alexandrine Egypt because conditions for its widespread use were nonexistent.

Industrial Capitalism intensified the global trends toward economic integration and heightened the surge of productivity started under Commercial Capitalism. By 1914 over 516,000 kilometers of cables had been laid on ocean beds, over 30,000 ships carried goods from port to port worldwide, and transcontinental railroads had opened up the United States, Canada, and Siberia for economic exploitation. Thanks to this evolving global infrastructure, the wool of New Zealand, the wheat of Canada, the meat of Argentina, the corn of the United States, the rice of Burma, the rubber of Malaya, the cotton of Egypt, the jute of Bengal, and the humming factories of western Europe and the United States were all enmeshed and coordinated in a dynamic global economy.

When something affected one element in this integrating economic system, the ramifications could be surprisingly widespread and complex. Between 1879 and 1902, for example, the cost of transporting eight bushels of wheat from Chicago to Liverpool dropped from an average of eleven shillings to below three shillings, due to improvements in oceanic shipping. The resulting flood of cheap New World wheat undercut European agriculture and stimulated a flood of European peasant emigrants to embark for the Americas.

As Industrial Capitalism gained momentum, the joint-stock companies that had conducted international business under Commercial Capitalism gave way to a new form—the cartel or combine. The American steel magnate, Andrew Carnegie, noted that the watchword of every business had to be "expand or perish." About his own industry, he observed that "when the manufacturer of steel ceases to grow, he begins to decay, so we must keep extending." This inherent impulse led to the replacement of family concerns by partnerships and then by cartels, which dominated individual industries. The cartels extended their operations globally in order to gain economies of manufacturing, to control sources of raw materials and markets for their products, and to regulate prices so that profits could be ensured. Examples of these international cartels in the late nineteenth century were the Swedish Match Company, the Nobel Dynamite Trust Company, the Dunlop Rubber Company, Royal Dutch Petroleum, and Lever Brothers (which produced soap and margarine).

The cartelization of the global economy led to enhanced global economic productivity. Between 1860 and 1913, world trade increased twelve times over and world industrial production seven times. Since Europe had initiated and coordinated the new global economy, Europe was its main beneficiary. By the late nineteenth century, some of these incoming profits had trickled down to those at the bottom of European society. Whether the real wages of workers rose or fell during the early days of the Industrial Revolution remains a disputed issue. Undisputed though is the steady rise of real wages in the second half of the nineteenth century. Between 1850 and 1913, the real wages of British and French workers roughly doubled. At the same time, social-reform legislation adopted by all Western governments began to provide workers with such benefits as medical insurance, accident insurance, unemployment insurance, a reduced working day, and improved workplace conditions.

This general improvement in living standards was reflected in a dramatic threefold increase in Europe's population between 1750 and 1914 (despite the emigration of millions of Europeans during that period). As nutritional levels rose, natural resistance to disease rose, too, and mortality rates dropped. At the same time, the Industrial Revolution made possible improved sewage systems and safer water supplies, further lowering mortality rates. The advances achieved by public health measures were crowned by the discoveries of medical science. Bacteriology emerged as a full-fledged discipline whose practitioners traced particular diseases to particular microbes, and developed specific vaccines and antitoxins for coping with them. Celebrated scientists such as Louis Pasteur in Paris, Joseph Lister in London, and Robert Koch and Rudolf Virchow in Berlin were responsible for brilliant discoveries that unlocked the secrets of many dread diseases and provided the means to immunize whole populations against some of them. By the end of the nineteenth century, such scourges as cholera, plague, and typhoid were disappearing from the European world; progress was being made in the control of diphtheria and syphilis; and Lister's work in asepsis had greatly reduced the number of deaths resulting from childbearing and from general surgery.

The application of these medical discoveries and the routine segregation of infected persons further reduced European death rates from thirty per one thousand persons in 1800 to fifteen in 1914. Europe's population, in turn, rose from 188 million in 1800 to 463 million in 1914. At the same time, European emigrants settling Australia, Siberia, and the Americas pushed the percentage of people of European origin from roughly one-fifth to one-third of total world population. In this way, the new global economy created by Industrial Capitalism altered basically not just the world economic balance but also the world demographic balance.

The New Politics
During the centuries of Commercial Capitalism, the relatively integrated states organized by the European monarchs proved more efficient and aggressive than the empires of the Aztecs, the Ottomans, or the Moguls. This discrepancy in political power only increased with the coming of Industrial Capitalism, and resulted in the European empires that partitioned the globe into colonies or semicolonies.

One reason for this growing power gap was the accelerating development of European military technology, ranging from light, cheap field artillery that could be fired as rapidly as muskets, to steam-powered warships with heavy ordnance. As European armies were, for the first time, organized along industrial lines as machines of war, the gap between Western and non-Western military forces grew ever wider, far more so than several centuries earlier when the Amerindian empires crumbled before the conquistadors. As early as November 1839 when the Opium War began, it quickly became apparent that British warships could freely shell the Chinese coast with no feasible response. Confronted with the manifestly superior power of the Royal Navy, the Chinese General Staff could conceive of nothing better than a plan to tie firecrackers to the backs of monkeys, which were to be flung on board the British warships. The hope was that flames from the scrambling monkeys might reach the ships' powder magazines and blow them up. Nineteen monkeys were actually delivered to headquarters, but nobody could figure out how to get close enough to the British ships to throw the monkeys. So as the British squadron seized port after port, the unfortunate monkeys, deserted during the panic following the British attacks, died of starvation in their cages.

The widening power gap between West and non-West also reflected the greater economic dynamism of Industrial Capitalism. Rising living standards in western Europe stimulated popular demand for soap, margarine, chocolate, tea, cocoa, and rubber for bicycle tires. Large-scale imports from tropical regions necessitated the creation of local infrastructures of harbors, railways, highways, warehouses, and telegraph and postal systems. Such infrastructures required local regimes of order and security to ensure adequate dividends for shareholders back in the European homelands. Hence, the clamor for the annexation of faraway territories if local conflicts disrupted trade (as in Egypt in 1882 when the revolt of the nationalistic Colonel Ahmed Arabi prompted an invasion by the British to safeguard their extensive financial and commercial interests) or if a neighboring colonial power threatened to expand into an area considered within your sphere of economic or political primacy (as in Libya, invaded by Italy in 1911 to forestall the British in neighboring Egypt and the French in neighboring Tunis).

The possibility of encroachment by other powers became an obsessive concern as the Industrial Revolution spread and Germany and the United States with their newly built, ever more efficient plants threatened to overtake Britain as economic powerhouses. Although world industrial production increased seven times between 1860 and 1913, British production only tripled, and French production only quadrupled—impressive statistics in themselves, but not when placed against Germany's sevenfold leap and the United States' twelvefold one in the same period. The colonial repercussions of this economic rivalry were far-reaching. As the British author Sir John Keltie put the matter in 1895, not "until Germany came into the field ten years ago" were Britain's complacent capitalists moved "to look around and look forward." By that time, only Africa "remained available" for colonization and so "on Africa a rush was made without precedent in the history of the world."[21]

Typical was the British entrepreneur Sir George Goldie, who in 1879 organized the United African Company to compete with the French in the Niger Valley. "With old established markets closing to our manufactures, with India producing cotton fabrics not only for her own use but for export, it would be suicidal to abandon to a rival power the only remaining undeveloped opening for British goods."[22] To avert such "suicide," Goldie established over 100 trading posts in the Niger Valley, and backed them up with 237 treaties that his agents concluded with African chiefs, and that invariably ceded to the United African Company "the whole of the territories of the signatories" along with the right to exclude foreigners and to monopolize the trade of the involved territories. Thus, the company became the de facto government of the Nigerian hinterland well before the Berlin Conference of 1884–85 officially accepted the British government's claim to those territories.

Other European powers were also carving out their slices of the continental pie. In 1879 the only colonies in Africa had been those of France in Algeria and Senegal, of Britain along the Gold Coast and at the Cape, and of Portugal in Angola and Mozambique. By 1914 the only uncolonized remnants of Africa were Ethiopia and Liberia.

European empire building was fueled not only by overwhelming military and economic power, but also by a powerful psychological impulse. Social Darwinism, with its doctrine of survival of the fittest, provided a persuasive ideological underpinning for global political expansionism. While Rudyard Kipling wrote of the "white man's burden" in bringing the blessings of civilization to "lesser breeds," Cecil Rhodes, the exuberant British empire builder in southern Africa, unabashedly proclaimed that "we are the first race in the world and that the more of the world we inhabit the better it is for the human race."[23]

According to Rhodes's criterion, the world was indeed well off by 1914. All of Africa and India and most of Southeast Asia were European colonies; the Chinese, Persian, and Ottoman empires were semicolonies; and Latin America, an economic appendage of Britain in the nineteenth century, was a vast semicolony of the

United States in the twentieth. It is not surprising that Rhodes ended up dreaming of other planets to conquer. "The world is nearly parcelled out, and what there is left of it is being divided up, conquered, and colonized. To think of these stars that you see overhead at night, these vast worlds which we can never reach. I would annex the planets if I could; I often think of that. It makes me sad to see them so clear and yet so far."[24]

DESTRUCTION

Alexis de Tocqueville, the noted interpreter of early American democracy, left us a classic description of the basic contradicitions that lay within Industrial Capitalism when in 1835 he visited Manchester, Britain's first great textile manufacturing center, whose hundreds of mills churned out the fabrics that clothed tribespeople in Africa, plantation hands in the Antilles, and peasants in the Orient. Of this first urban legacy of the Industrial Revolution, de Tocqueville wrote: "From this foul drain the greatest stream of human industry flows out to fertilize the whole world. From this filthy sewer pure gold flows. Here humanity attains its most complete development and its most brutish; here civilization works its miracles, and civilized man is turned back almost into a savage."[25]

As with Commercial Capitalism, the destructive impact of Industrial Capitalism was most evident and most thoroughgoing in those vulnerable colonies or dependencies far from Europe. Now, however, its ravages were not confined to the Americas and scattered regions of Africa, but affected all lands. Indeed, the classic example of capitalism's impact in this period was India, where the English, unlike the many other previous conquerors of the subcontinent, wanted not just booty and tribute, but above all else a free field for their enterprise, an enterprise that turned India's traditional economy and society inside out and upside down.

First, the British wanted Indian raw materials such as jute, oilseeds, wheat, and cotton. To transport these commodities, they had built seven thousand miles of railroad in India by 1880, and dug the Suez Canal to reduce the shipping distance between England and India by over 40 percent. The same railroads that carried away India's commercial crops brought back cheap, British machine-made industrial products to its villages.

The inexpensiveness of these imports undermined India's crafts, especially that of its textile artisans, who had formerly produced the most desired cloth in the world. For the second part of what the British wanted from India was that it become an enormous market for its industrial goods, especially its textiles. The British began their textile offensive in a defensive manner by levying a 70 to 80 percent duty on Indian cotton textiles imported into Britain. This they did in 1814, while they were still perfecting their new cotton machines.

The prospect of global markets stimulated a rapid succession of mechanical inventions that in a few decades made British textiles more than competitive with Indian ones in quality and price. When this happened, the British promptly swamped India with their cheap, machine-made textiles, and forbade import duties that might have protected Indian industry as the British had been protected earlier in the century.

Between 1814 and 1844, the number of pieces of Indian cotton goods imported into Britain fell from 1.25 million to 63,000, while British cotton exports to India rose from 1 million to over 53 million yards. The impact on India's ancient textile crafts was shattering. While Manchester rose to world prominence and increased in population from 24,000 in 1773 to 250,000 in 1851, Sir Charles Trevelyan testified before a parliamentary committee in 1840 that "Dacca, the Manchester of India . . . has fallen from 150,000 to 30,000, and the jungle and malaria are fast encroaching upon the town."[26]

Meanwhile, Western medical science, health measures, and famine-relief arrangements helped boost India's overall population from an estimated 255 million in 1872 to 305 million in 1921. Europe had, of course, experienced a similar population increase, but its developing factory system was ready to absorb new workers, and emigration to the Americas and Australia provided an additional safety valve. No such outlets were available to Indians (and other overcrowded colonials) because Britain actively discouraged Indian industries, and because the "empty" overseas lands had been preempted by European colonists who promptly erected immigration barriers against the "lesser races." India's extra millions had little choice but to remain in their villages, creating a terrible crisis of rural overpopulation in relation to the resources the countryside had available. This remains to the present day one of the most acute problems of the Indian economy, and indeed of most Third World economies. As elsewhere, the underlying factor was the disruptive, if varying, impact of Western Industrial Capitalism on traditional societies—more intensive on the island of Java than on the subcontinent of India, and more pervasive in colonized India than in China, whose coastal ports passed under Western control, but whose vast interior retained considerable autonomy.

Industrial Capitalism and the New World imperial order disrupted not only traditional economies but also traditional cultures, thoroughly penetrating and undermining the confidence of local elites. This thin, educated upper crust in the newly colonized or near-colonized world beyond Europe had tended initially to look down on Europeans as uncouth barbarians. Even as late as the 1890s, a Chinese scholar could write his brother these impressions of Westerners: "Their eyes have a peculiar look in them; they lie on a straight line, and are green, and blue and sometimes brown. Their cheeks are white and hollow, though occasionally purple; their noses like sharp beaks, which we consider unfavorable. Some of them have thick tufts of hair, red and yellow, on their faces, making them look like monkeys. Their arms and ears do not reach to the ground as they are depicted by us. Though sleepy, I think they have intelligence."[27]

However, the undeniable military might, and the scientific and technological achievements that lay behind Europe's imperial thrust, along with the stunning wealth of the European countries, soon led many members of non-European elites not just to learn a Western language and to read Western newspapers and books, but to become uncritically admiring of all things Western in their search for the key to the West's awesome "power and wealth." The human results of this shift from thoughtless disparagement to equally thoughtless deference were noted by the prominent Indian nationalist leader, Surendranath Banerjea:

Our fore fathers, the firstfruits of English education, were violently pro-British. They could see no flaw in the civilization or the culture of the West. They were charmed by its novelty and its strangeness. The enfranchisement of the individual, the substitution of the right of private judgement in the place of traditional authority, the exaltation of duty over custom, all came with a force and suddenness of a revelation to an Oriental people who knew no more binding obliga-

tion than the mandate of immemorial usage and of venerable tradition. . . . Everything English was good—even the drinking of brandy
was a virtue; everything not English was to be viewed with suspicion.[28]

Quite different were the reactions of the illiterate peasant masses
to the West's intrusion. They tended to idealize their traditional
village communities, now disintegrating, which had once provided
land for all their members, preserved the continuity of interpersonal
relationships, and assured a sense of individual status in society. Of
course, they had also been vulnerable to war, famine, pestilence,
and every sort of despotic arbitrariness. Karl Marx in an 1853 dispatch to the *New York Daily Tribune* castigated British abuses in
India, but then added: "I share not the opinion of those who believe
in a golden age of Hindostan. . . . We must not forget that these little
communities were contaminated by distinctions of caste and by
slavery."

Marx's disclaimer was fully justified, but in the same dispatch he
emphasized the difference between exploitation under tributary relations and exploitation under capitalism. The latter, wrote Marx,
"has broken down the entire framework of Indian society, without
any symptoms of reconstitution yet appearing. This loss of his old
world, with no gain of a new one, imparts a kind of melancholy to
the present misery of the Hindoo, and separates Hindostan ruled
by Britain, from all its ancient traditions, and from the whole of its
past history."[29] More specifically, the "melancholy" to which Marx
referred arose from the new legal, administrative, and security
systems which the British imposed on Indian rural society to enforce and control the needs of the new monetized society they had
created. In the new British India, "land" became a profit-yielding
possession; "food" a mere commodity of exchange; "neighbor" a
common property owner; and "labor" a means of survival. Such was
the essence of capitalist "destruction" in the colonial world.

Another commonly overlooked aspect of this "destruction" was
the Asian contract coolie labor that Industrial Capitalism harnessed
for its global plantations, just as in an earlier period Commercial
Capitalism had harnessed African slave labor for its New World
plantations. Improved ocean transportation and the opening of the
Suez Canal in 1869 suddenly made it commercially feasible to ship

plantation produce from Asia to western Europe. Commodities such as rubber, tea, coffee, pineapples, and cocoa were now produced on European-owned plantations in southern and East Africa, South and Southeast Asia and the islands of the Caribbean and the Indian and Pacific oceans for the European market.

Since slavery had been abolished by Western governmental decrees in the mid nineteenth century, the new plantations needed to find a new type of labor supply. This was provided by "coolies," as the unskilled Asian laborers were commonly called. Mostly from China and India, they accepted "contracts" providing for long-term indenture and penal sanctions in cases of nonfulfillment. For the illiterate coolies, transported far from their native villages, these contracts usually provided little protection against exploitation, either because they were not honored, or because oral promises made by recruiters were not actually included in the contract. In this manner, more Asian peasants were recruited as laborers for the new global plantations than Africans had been dragooned into slavery in the Americas in earlier centuries. Despite the large number of coolies that left their native lands, the pressure of overpopulation in Asian villages was not alleviated. The population base in China and India was so enormous that the emigration, massive though it was, nevertheless did not constitute a significant safety valve.

Although Industrial Capitalism's destructive tendencies were most obvious in non-European communities, Europe was by no means spared. There the peasants and artisans, rather than being trapped in overcrowded villages, were uprooted and forced to cope with the trauma of moving to new factory towns, finding employment, and adjusting to strange new ways of working and living. Having no land, no cottages, and no tools, they had nothing to offer but their labor. They had become mere wage earners, completely dependent on their employers. The monetization of English society was as wrenching for English have-nots as the monetization of Indian society was for the Indian peasantry.

From its very beginning, the new industrial society of England was both praised and deplored for its effects on the general population. Supporters argued that real wages were higher in factory towns than they had been in the villages, which explained the steady supply of urban labor that flowed out of both the English and Irish countrysides. Detractors stressed the unwholesome and dangerous conditions in factories and tenements, the exploitation of child and female labor, the rigid discipline imposed by ever-present overseers, and the long hours and deadly monotony of keeping pace with the machines and adapting daily life to a factory whistle rather than to the rising and setting of the sun.

Whatever the merits of the controversy, there can be no question about the intensity of the workers' discontent. They burned Arkwright's factories in Lancashire and smashed Hargreaves's spinning jennies at Blackburn. Historian Asa Briggs concludes that England, after defeating Napoleon's armies in 1814, came closer to social revolution than at any time in its history. The authorities responded by making machinery wrecking a capital offense and by shipping thousands of militant workers to Australia. Britain's Labour Party leader Aneurin Bevan concluded a century later: "It is highly doubtful whether the achievements of the Industrial Revolution would have been permitted if the franchise had been universal. It is very doubtful because a great deal of the capital aggregations that we are at present enjoying are the results of the wages that our fathers went without."[30]

Mass suffering and resentment stimulated numerous working-class self-help movements, including friendly societies, trade unions, mechanics' institutes, cooperatives, and reading societies where literate workers read aloud from newspapers and tracts to their illiterate companions, and even taught some to read and write. Among the middle and upper classes, the Industrial Revolution spawned two critical responses. The conservative one, voiced by Edmund Burke in England and Joseph de Maistre in France—"prophets of the past" as a wit dubbed them at the time—looked back nostalgically to preindustrial rural communities where everybody knew everybody else, and where the impersonality of the city was mercifully unknown. Their ideal society was aristocratic but not plutocratic, socially responsible rather than irresponsible, and promoted upper-class paternalism not laissez-faire social neglect.

Whereas conservatives held that humans would degenerate to an animallike state without the restraints of traditional religion and social order, the radicals looked to the future, believing human nature to be primarily the product of social environment. Therefore, they favored a basic restructuring of society to promote collective well-being and cooperative behavior. They deplored the social dis-

126

ruption and anomie associated with industrialization as much as the conservatives did, but instead of seeking redress by restoring the past, they proposed sweeping social innovation. Their various specific solutions to what they viewed as the ills of industrial society comprise the rich assortment of "isms" that distinguish our own century—anarchism, Christian socialism, utilitarianism, syndicalism, utopian socialism, and Marxian socialism.

Marxism eventually became the most influential of these isms, partly because its theory of class struggle held that the socialist goal would be won by the workers themselves, and that their eventual victory was assured because capitalism bore within itself "the seeds of its own destruction." Recurring economic crises, rising unemployment and falling living standards in the European heartland of capitalism would force exploited proletarians to take up arms, overthrow capitalism, and finally establish their own socialist order.

Events, as it turned out, took quite a different course. Living standards rose rather than fell, and workers gradually won the right to vote. The ballot called the need for the bullet—at least as an instrument of internal revolt—into question in western Europe and the United States. This unexpected denouement divided the Marxists into warring factions. Orthodox socialists clung to their original class-struggle doctrine, but by the eve of World War I the majority of western European workers were backing "revisionist" socialists whose slogan was "Work Less for the Better Future and More for the Better Present."

THE TIME OF TROUBLES, 1914-40

The "better present" failed to materialize. Instead, the first part of the twentieth century proved to be a time of troubles for world capitalism, which experienced three major setbacks: the trauma of World War I, the breakaway and consolidation of the USSR, and the ravages of the Great Depression.

The literature on the origins of the First World War is enormous. Many factors were involved, including the psychological influence of social Darwinism's "struggle for survival," and the nationalist aspirations of Europe's numerous subject minorities, which provided the spark at Sarajevo (where a Serbian nationalist assassinated the Hapsburg Archduke Francis Ferdinand) that set off the war and then tore to pieces the multinational empires of central and eastern Europe. Industrial capitalism also contributed substantially to the outbreak of war as newly industrializing states with more modern and efficient plants overtook the industrial pioneers and generated a fierce struggle for dominance of the emerging global economy. Thus, Britain in 1870 produced 31.8 percent of world industrial output; Germany only 13.2 percent. By 1914 Britain's share had dropped to 14 percent while Germany's had risen slightly to 14.3 percent.

Economic competition involving numerous powers generated a naval armament race—the precursor of our own nuclear arms race today—as strong navies came to be considered essential for protecting imperial trade routes, and also sparked a race for colonies, prized for their raw materials and as potential markets for surplus capital and manufactured products. Ever present, if sometimes in the background, was the psychological goad of imperial prestige, the resolve of all the Great Powers to stake out and possess what they considered to be their rightful "place in the sun." Hence, the scramble for empire in the late nineteenth century and a series of recurring imperial conflicts throughout the world: Britain versus Germany in East Africa and Southwest Africa; Britain versus France in Siam and the Nile valley; Britain versus Russia in Persia and Afghanistan; and Germany versus France in Morocco and West Africa.

When the First World War erupted in the summer of 1914, the British foreign secretary, Sir Edward Grey, remarked that "the lights are going out all over Europe." His comment proved remarkably prophetic. Only four years later, the centuries-old Hapsburg, Hohenzollern, Romanov, and Ottoman empires had disappeared replaced by new institutions, new leaders, and new ideologies that aristocrats like Sir Edward could not have foreseen, no less comprehended.

World War I differed from all previous wars in its degree of mass mobilization, involving not only combatants but civilians, not only a material mobilization but an intellectual one. The mustering of enormous armies, totaling tens of millions of men, necessitated a corresponding civilian mobilization. With compulsory universal military service for all men, there was an informal mobilization of women to fill their places in industry and agriculture. Chronic shortages in war matériel of all types forced governments to mobilize industry, leading to what was called "war socialism." Human minds were also "mobilized," as all governments waged a war of propaganda depicting the conflict as a stark struggle between the purest of good and evil, with God recruited as a fighting compatriot by all belligerents. World War I was unprecedented to that time also in the deadliness of its weapons, in the number of casualties sustained, and in the extent of material devastation endured. From the opening campaigns, machine guns proved their efficiency in mowing down advancing infantrymen like blades of grass. As the war continued, a succession of new weapons was introduced: poison gas, tanks, flamethrowers, submarines that torpedoed all ships in enemy waters, and airplanes that began dropping bombs on open cities. By war's end, 8.5 million combatants and 10 million noncombatants were left dead, and huge swaths of Europe were in ruins.

The unprecedented destructiveness of the First World War enabled the United States to leapfrog over the European powers and into the position of the number-one financial and industrial power of the globe. In the colonial world, the tribulations of the imperial powers undermined the idea of "white men" as divinely ordained rulers. The wartime promises of the Allied Powers (and President Woodrow Wilson in particular) raised hopes among subject peoples everywhere, and led young nationalists in countries like China, India, Vietnam, and Egypt to demand that the much-touted principle of "self-determination" be applied to Africans and Asians as well as Poles and Serbs. Though such hopes were soon dashed by the imperial peace that followed, the seeds of a new nationalist resistance in the colonial world had begun to germinate and would come to fruition after the next great war.

Finally, World War I so devastated Russia and its antiquated military that it cleared the ground for the 1917 Bolshevik revolution, which led to the first major defection from the ranks of the capitalist states. Lenin's slogan during the war years had been that workers and peasants in all countries should "turn the imperialist war into a class war." After experiencing years of unprecedented bloodshed and destruction, Russia's peasants and workers, who originally had supported the war against the hated Germans, responded to the militancy of Lenin's appeal and in October 1917 succeeded in overthrowing the short-lived provisional government that had supplanted the tsarist regime. Lenin then declared the establishment of a soviet state dedicated to creating a socialist society in Russia and advancing the cause of socialism abroad.

Winston Churchill once declared that "the failure to strangle Bolshevism at its birth ... lies heavily upon us today."[31] It was not, however, due to any lack of effort on the part of the capitalist West (and Japan) that the infant Bolshevik state escaped strangulation. Under Churchill's leadership, the Allies, after Germany's surrender in November 1918, sent interventionist armies to attack the beleaguered Bolsheviks—through the Ukraine in the South, Poland and the Baltic region in the West, Murmansk in the North, and Vladivostok in the East. The overall strategy was to assist the anti-Soviet forces within Russia, led mostly by former tsarist generals and landowners, to overthrow the Bolshevik regime. The latter managed to survive largely because most Russian peasants believed that under the Bolsheviks they would have a better chance of keeping the lands of the nobility and the church that they had seized during the revolution. With the departure of Japanese forces from Siberia in 1922, the years of bloodshed and destruction finally ended, and the future of the Union of Soviet Socialist Republics was ensured.

From the viewpoint of the capitalist West, as threatening as the failure to "strangle" Bolshevism at birth was the subsequent success of the Bolsheviks in building a planned economy. It was assumed by many that the Soviet experiment was bound to fail because it was contrary to "human nature." Lenin himself admitted that the Bolsheviks had no specific blueprint for the building of socialism when he admitted, "We must go by experiments."[32] In addition, the new regime faced almost inconceivable problems. The world war, the revolution, and the civil war that followed had, in the phrase of historian Moshe Lewin, "archaicized" Russia. What capitalist experiments had taken place in tsarist Russia before 1914 had been quite literally "peeled away" by ensuing events. The peasants had taken over the large estates of the landlords and returned them to a far more primitive type of family farming that left little agricultural surplus for the state to purchase or even commandeer. By the early 1920s, Russia's urban population had returned only to its prewar share of the total population. In short, postrevolutionary Russia was, in Lewin's words, "even more rural, and equally—if not more—backward than tsarist Russia."[33]

Lenin died in 1924, so it was his revolutionary successors who did the experimenting under far less than optimal conditions and against the background of a populace and a bureaucracy disoriented by the scope of the events they had experienced and, for the most part, lacking even the simplest levels of technical, educational, or cultural accomplishment. It was hardly surprising, then, that the country and the revolutionary experiment itself should be taken over at all levels by despotic rule. Even less surprising has been the repetition of this dismal pattern in rebellious Third World countries after World War II. Beginning in most cases at an economic level even lower than that of tsarist Russia, and subject to intense military, political, and economic pressures from the capitalist world, the failure of today's Third World countries to attain economic as well as political independence—and the spread of various forms of "revolutionary" despotisms to most of them—becomes understandable.

After much infighting within the Bolshevik Party, by 1927 Joseph Stalin had emerged alone as an all-powerful dictator and had decided in favor of a managed economy based on centrally prepared Five-Year Plans. The wisdom of this decision is now widely questioned within the Soviet Union as well as abroad, but the Five-Year Plans did produce quick and dramatic results by concentrating labor and material resources on the production of capital goods rather than consumer goods. By 1932, the year the first Five-Year Plan ended, the Soviet Union had risen in industrial output from fifth to second place in the world, or put differently, its share of total global industrial output had risen from 1.5 percent in 1921 to 10 percent in 1939.

This extraordinary leap was due not only to increased Soviet productivity, but also to the slump of global capitalism with the Great Depression. The United States stock market crashed in October 1929. Within one month, stock market values had dropped 40 percent, a decline that continued for the next several years. During those years, five thousand American banks failed, 13 million workers lost their jobs, and the U.S. steel industry operated at only 12 percent of capacity. Excluding the Soviet Union, world industrial production fell 36.2 percent between 1929 and 1932 (compared to an average drop of 7 percent in previous depressions), while international trade dropped by about 65 percent (compared to a maximum drop of 7 percent in past crises).

The contrast between a booming Soviet economy and a prostrate global capitalist economy made a deep impression on many, both in the industrialized and in the colonized worlds. India's future leader, Jawaharlal Nehru, for instance, wrote from his cell in a British jail: "While the rest of the world was in the grip of the depression and going backward in some ways, in the Soviet country a great new world was being built up before our eyes. . . . The great world crisis and slump seemed to justify the Marxist analysis."[34] In a similar mood, British historian Arnold J. Toynbee wrote: "In 1931, men and women all over the world were seriously contemplating and frankly discussing the possibility that the Western system of Society might break down and cease to work. . . . the members of this great and ancient and hitherto triumphant society were asking themselves whether the secular process of Western life and growth might conceivably be coming to an end in their day. . . ."[35]

Capitalism's time of troubles persisted through the 1930s. Apart from a few brief moments of partial recovery, unemployment continued at unprecedented levels, factories remained closed, and crops were left rotting in fields, while startling numbers of citizens lacked adequate food, clothing, and shelter. The question that both Nehru and Toynbee had raised about the future of capitalism remained one of urgent concern until the diplomatic crises of the late 1930s and subsequent preparations for war once again started factories humming and farmers working their fields. It is ironic as well as significant that capitalism's time of troubles was ended only by preparations for World War II, a war that in addition to resolving the prolonged capitalist crisis of the 1930s soon devastated much of the industrialized world (including the Soviet Union and Japan), while at the same time laying the groundwork for the technological leap forward that heralded the third phase of global capitalism—High-Tech Capitalism.

HIGH-TECH CAPITALISM, 1940 AND AFTER

In 1966, British scientist C. P. Snow said that "current inventions comprise the biggest technological revolution men have ever known, far more intimate in the tone of our daily lives, and of course far quicker, either than the agricultural transformation in neolithic times or the early industrial revolution."[36] In this statement, Snow captured the essence of the current high-tech revolution—its unprecedented tempo and potency. Equally unprecedented are the attendant problems this new phase of capitalism has generated, problems of such scope that there is now serious concern not merely about the "Western system of society," not just about human civilization, but about the survival of the human species itself. The "creative destruction" inherent in capitalism has now been raised to the nth power, forcing on us all a reappraisal of hitherto sacrosanct institutions and values. Evidence of the beginning of such a reappraisal can be found in the wide range of often inchoate, sometimes contradictory social innovations and experiments being tested around the world. That such social ferment has been largely ineffectual in the industrialized heartland is evident from the spectacle of rival social systems beset by such basic failings that each seeks legitimacy not in its own record but in the defects of competing ideologies, and in the Third World by the fact that some of its despairing citizens now are beginning to talk about the "good old days" of colonial rule.

CREATIVITY

Today's high technology is a historically unique development because of its military origins. Throughout history, preparation for war has stimulated technology, whether the flamethrowers of ancient Greece, the cannon and firearms of early modern western Europe, or the planes, tanks, and poison gases of World War I. In the period just before and during World War II, however, a qualitative change occurred in the relationship between the military, technology, and society. The locus for technological advance shifted decisively to the military—and later in the United States and the Soviet Union to what came to be called the military-industrial complex. Henceforth, technological advances were initially military in nature and only later "civilianized" as spin-offs from military-directed research and development.

An explosion on the desert floor of Alamogordo, New Mexico, on July 16, 1945, signaled the harnessing of the power of the atom. This power, first developed for military purposes and dropped to horrifying effect on Hiroshima and Nagasaki, is today used for many civilian purposes, including biomedical research, medical diagnosis and therapy, and the generation of electrical power. Likewise, the V-2 missiles the Germans developed to bombard London led within decades to weather satellites, telecommunication satellites, and planning for manned landings on Mars. No one can foresee precisely what the repercussions of this breakthrough into space will be, any more than Columbus's contemporaries were able to foresee the consequences of his discovery of the New World. But manufacturers of crystals, vaccines, and pure-tissue cultures are already exploiting opportunities in space offered by the absence of gravity, by a limitless access to vacuum conditions, and by the availability of superhigh and superlow temperatures to experiment in the creation of astounding new products. Futuristic plans are also on the boards for the industrial exploitation of deep space, and the construction of huge orbiting platforms or "islands" where human colonies might be established as they were in overseas lands a few centuries ago.

Early research on the computer, organized by the military as part of the war effort, led to various applications of the first crude computers before the end of the war. This general field advanced rapidly in the postwar years with the development of microconductors or silicon chips that allow machines to become both faster and smaller, and have now become the backbone of modern economies, being used in power stations, business offices, supermarket checkout stands, telephone switching systems, and factory production lines.

Genetic engineering, potentially a key element in the future development of High-Tech Capitalism, is distinctive because it emerged out of the civilian sector of the economy. The advances to come were triggered in 1953 by James D. Watson and Francis H. C. Crick's discovery of the structure of DNA (deoxyribonucleic acid), the chemical that carries the genetic code of all living things. With this knowledge, scientists could begin to learn how first to "read" genetic codes, then later to modify them, and most recently to take the first steps toward creating new ones. Instead of modifying plants and animals to human needs through generations of careful crossbreeding, as had been done by peasants and herders for millennia, scientists can now begin to choose among individual genes, manipulate them directly and even clone improved life forms. Such genetic engineering has already led to the creation of insulin and growth hormones in the laboratories, as well as several new vaccines, including one against the highly infectious hoof-and-mouth disease among cattle.

Genetic engineering is changing the nature of agriculture as well as medicine. The steep rise in the demand for, and the prices of, farm products during World War II stimulated the "green revolution." This consisted of new hybrid varieties of major cereals that greatly increased crop yields when used in conjunction with irrigation, fertilizers, and pesticides. In the 1980s, genetic engineering began to create a new-style green revolution by "splicing" together combinations of genetic material to create new life forms better suited to human needs. Scientists now foresee genetic engineering that will enhance the growth rates and disease resistance of animals and plants. They also foresee new plants that will grow in cold climates, in salty or dry soils, that make their own hydrogen fertilizers, and that are resistant to diseases caused by viruses, bacteria, fungi, and worms.

The high technology spawned by World War II helped power the quarter-century boom after the war, a golden age of global capitalism. In addition, the boom was strengthened by cheap oil, which literally fueled the industrialized economies of the First World at negligible cost. Before OPEC (Organization of Petroleum Exporting Countries) was formed in 1960, the Western-controlled international oil companies were free to set the price of oil as low as $1.20 a barrel—a giveaway that represented a crucial stimulant to economic growth. Finally, the global boom was initially driven by the need to repair massive war damage, by a pent-up demand for consumer goods neglected during the war years, and by vast military purchases from the Korean War through the Vietnam War. The net result was that between 1955 and 1980, world output (the sum total of all countries' GNP) tripled in real terms, measured in constant dollars, and during that same quarter century, global GNP per capita doubled, even though the world population rose from 2.8 to 4.4 billion people.

Multinational corporations spearheaded this global economic expansion, playing the same lead role under High-Tech Capitalism that joint-stock companies had under Commercial Capitalism, and cartels under Industrial Capitalism. These multinationals have been able to integrate the global economy to an unprecedented degree, and often without regard to national sovereignty, because of the new technological facilities available to them. These include containerized shipping that lowers transportation costs in a way that permits the exportation not just of manufactured products but of production machinery or even entire factories; improved engineering techniques making possible the breakdown of labor, production, and assembly operations into a series of processes that can take place thousands of miles apart and hence can take full advantage of the cheapest local unskilled, semiskilled, and skilled labor anywhere on the globe; and the combining of computers and satellite communications for the instantaneous global coordination of production.

Precisely what this means may be illustrated by comparing the operations of the steel magnate Andrew Carnegie at the end of the nineteenth century with the operations of multinational corporations at the end of the twentieth century. Carnegie justifiably boasted that his steel company mined iron ore on the shores of Lake Superior, transported it nine hundred miles to his mills in Pittsburgh, combined it with coke, lime, and manganese shipped in from other parts of the country, and ended up with steel costing only two cents a pound. Today the multinationals carry on their activities on such a larger scale that they differ from Carnegie's steel company as it differed from a colonial blacksmith in the Thirteen Colonies. Their technology incorporates electronics, genetic engineering, satellite communications, and automated factories and offices—none of which were either known or imagined in Carnegie's time. Multinationals also operate at a new financial level, using computerized and centralized cash-management systems, tapping world money markets, and arranging to have their accounts payable in weak currencies and receivable in strong currencies. Likewise, the worldwide dispersal of their factories enables multinationals to organize for the first time a truly global division of labor, enabling them to play off their expensive unionized labor forces at home against the cheap regimented labor available in Third World countries. Multinationals have also opened new frontiers in marketing as well as in production. Multibillion-dollar advertising campaigns together with the shaping of values and tastes by radio and television networks, and by movies, records, magazines, and comics, have all combined to create a global marketplace literally inconceivable to Carnegie.

This unique synchronization of historical forces explains why the median multinational corporation today is a global enterprise producing twenty-two products or partial products in eleven different countries. During the quarter-century post–World War II boom, American multinationals alone grew at an average rate of 10 percent a year, while purely national corporations in the United States grew at 4 percent a year. By 1980 between one-quarter and one-third of the world's total industrial output was produced by multinationals, and about half of all world trade consisted of transactions among multinationals.

As it had under Industrial Capitalism, so in this world of multinational high-tech capitalist prosperity some wealth trickled down to the blue-collar level of society—at least in the countries of what had come to be called the First World. This post–World War II improvement was achieved at an incomparably higher level than that of a century earlier, with workers in heavy industries such as mining, steel, and autos receiving wages that enabled them to purchase the trappings of a "middle-class" life-style. After basic needs were met, enough money was left over for weekend outings, annual vacation trips, and, even if cash was not in hand, the purchasing on credit of private homes, cars, and other durable goods. Many economists were so impressed by the effects of this quarter-century boom that they persuaded themselves their neo-Keynesian strategies had ensured purchasing power adequate to avoid the boom-and-bust business cycles of the past. This optimism was especially evident in the United States, where publisher Henry Luce was heralding what he dubbed the "American Century," and sociologists were proclaiming the coming of a "postindustrial society" in which the greatest human problems would revolve around issues of personal fulfillment.

By the 1980s, the notions of an American Century, depression-proof economics, and postindustrial society seemed almost as quaint and unrealistic as Cecil Rhodes's vision of a world map colored mostly British red. The assumption of continuous and accelerating economic growth had begun to be tempered by a perception of endemic maldevelopment as manifested in environmental degradation, resource depletion, institutionalized runaway consumerism, chronic unemployment, a crisis of values and cultural identities, and social afflictions such as homelessness, family instability, and drug abuse. An acclaimed stocktaking of the state of American society, published in 1986 under the title *Habits of the Heart* concludes that "unless we begin to repair the damage to our social ecology, we will destroy ourselves long before natural ecological disaster has time to be realized."[37] On the two hundredth birthday of the American Constitution, the cover of the magazine *Economist* of London depicted a frowning Statue of Liberty, captioned: "Whatever happened to America's smile?"

What is so striking about the world of the late twentieth century is the absence not only of America's smile, but also of those sunny heroic facades so familiar in the socialist societies of the Second World. Despite their professed commitment to Marxist principles, socialist societies everywhere have spawned new ruling elites, strengthened rather than dismantled the state, failed to develop self-reliant economies independent of global capitalism, and manifested many of the social disorders that they once associated with the "degenerate" West. Certainly, Secretary Mikhail Gorbachev's critique of Soviet society in his book *Perestroika* is at least as severe as the one offered of United States society by several American sociologists in *Habits of the Heart.*

Even more unsettling, given the expectations they roused in the 1950s and 1960s, is the current plight of the countries of the Third World that began the postwar era with a series of stunning political triumphs that resulted in the dismantling of the European empires. Now, after decades of social experiments and sacrifices, these countries are discovering that "development" and an increased GNP may only aggravate rather than alleviate social inequity and human misery. In some Third World regions, the quality of life has actually been deteriorating in recent years despite all the plans and activities during the successive "Development Decades" sponsored by the United Nations.

The essential quality of the 1980s is that *all* societies are spinning their wheels. The root cause of this global malaise is that all societies are experiencing, with greater intensity than at any time in the past, the penalties as well as the rewards of capitalism's creative destruction.

DESTRUCTION

The Third World

As in the past, it is the ex-colonial lands that have been the most vulnerable to the destructive dynamics of capitalism. The peoples of these lands began auspiciously with striking political victories in the immediate postwar years. Within a decade of World War II, virtually all the Asian colonies had won their independence, and a decade later virtually all the African colonies had followed suit.

Such political triumphs were followed, however, by economic disasters in the newly liberated countries. Paradoxically enough, these economic disasters were facilitated by the winning of political independence. So long as colonial rule prevailed, the colonized lands were shielded to a certain degree from the full blast of the global capitalist economy and of its global rivalries. But when the colonies broke away from their mother countries, they found themselves alone and vulnerable to the economic pressures of multinational corporations and to the political machinations of Great Power policy-makers who viewed the weak former colonies as pawns to be moved about on a global political and economic chessboard. Historian Theodore Von Laue has observed that "liberation meant a far more intensive submission to the Western model." So long as colonial rule prevailed, "it preserved a semblance of indigenous authority even while transforming indigenous society; it provided internal peace and order plus protection from the pressures of global politics."[38]

With the removal of the imperial cushion, jarring external intrusions have disrupted all phases of life in the Third World. Peasants have been uprooted by the new global system of high-technology production as pitilessly as European peasants were by the enclosures centuries earlier. While the high-tech green revolution has led to increased crop productivity in many Third World countries, only medium- and large-scale farmers have benefited because only they possess the capital needed for hybrid seeds, special chemical fertilizers, irrigation systems, and other complex and expensive equipment. As these larger farmers have introduced labor-saving machinery like tractors and harvesters into the fields in search of further profits, already underemployed peasants have been forced off the land and into city slums in search of work. There, in the *favelas* of Brazil, the *ranchos* of Argentina, and the *bidonvilles* of North Africa (so-called after the *bidons* or metal oil drums used in constructing the shanties in which the impoverished peasants live), they find themselves as underemployed and superfluous as they had become in their villages.

Mexico is typical of this process. Large-scale farmers there are increasingly using their fields to grow winter fruits and vegetables for the North American market, while the country is forced to import corn and beans from the United States as staples to feed its people. The number of landless Mexican peasants has risen from 1.5 million in 1950 to 5 million in 1980, many of them crossing the border illegally into the United States in search of work, or internally emigrating to Mexico City, whose population ballooned to an almost unmanageable 14 million by 1980 (with an additional 14 million expected by the year 2000).

The tragedy of this runaway urbanization, under way on all continents, is that few Third World countries have an urban industrial base capable of absorbing and utilizing even a fraction of these peasants. Newcomers to the cities are condemned to waste their lives trying to subsist on such work as street vending, shining shoes, running errands, pushing carts, or pedaling rickshaws.

High-Tech Capitalism, by so fully controlling the global production and marketing process, has placed each individual Third World nation producing foodstuffs and raw materials at a distinct and increasingly severe disadvantage. It has done this and devastated Third World economies to a large extent by forcing to historically rock-bottom levels the prices of their agricultural and raw-material exports in relation to the prices of the manufactured goods and the luxury goods they import from the First World. Despite the fact that famine and malnourishment stalk significant portions of the Third World, the global output of foodstuffs since World War II has actually been substantially greater than the increase in global population, which in the world markets has depressed crop prices. In other words, even when Third World countries "succeeded" by First World standards—by applying green revolution advances to crop technology to produce more food—the result at a global level is simply a drop in prices and further impoverishment. The prices of raw materials like Chile's copper, Kenya's coffee, and Bangladesh's jute have slumped sharply, too, because thanks to technological advances, the amounts of such raw materials needed per unit of industrial production are now only 40 percent of what they were in 1900.

With fiberglass cable replacing copper wire in the transmission of telephone messages, with plastics replacing steel in automobile bodies, and with an accelerating number of similar substitutions in other areas of industrial production, economist Peter F. Drucker has concluded that "foodstuffs and raw materials are in permanent oversupply" and "it is quite unlikely that [their] prices will ever rise substantially as compared to the prices of manufactured goods ... except in the event of major prolonged war.... The raw material economy has thus come uncoupled from the industrial economy. This is a major structural change in the world economy."[39] It is a change that raises basic questions about whether Third World countries will ever be able to repay their astronomical First World debts. How will such countries ever find the surplus necessary to feed and clothe their own populations while fulfilling any "modernizing" dreams, if they cannot raise investment capital by selling what goods they produce? Is it possible even to conceive of global peace and economic stability as long as the majority of the human race remains mired in such hopeless—and increasing—pauperization and consequent turmoil?

High-Tech Capitalism has also had the effect of distorting the industrial development of the Third World. During colonial times, local industries were normally actively discouraged by the imperial powers, which wanted their colonies to remain only consumers of manufactured goods. After World War II, the newly-independent countries tried to redress their economic backwardness by import-substitution industrialization—that is, by helping local industries to manufacture formerly imported goods through such measures as protective tariffs, cheap credits, cheap energy, and the provision of adequate infrastructure facilities, including transportation and communications systems. This strategy enjoyed some success during the global boom of the 1950s and 1960s, but as the boom withered, so did the local industries. They did so primarily because poverty-stricken peasants lacked the purchasing power to buy their products, while the wealthy elite who did have money were too few in number to support such industrialization as consumers; and in any case preferred to buy fashionable luxury goods imported from abroad. .

This failed economic strategy was then replaced in the 1960s in many Third World countries by a strategy of export-directed industrialization in which the multinational corporations rather than local industries were offered centrality of place. The multinationals received very generous concessions from Third World governments, including cheap or free factory sites, inexpensive infrastructure facilities, preferential tax treatment, unrestricted repatriation of profits, and a cheap, regimented labor supply. All this made possible the turning of a number of Third World countries into "export platforms"—that is, extensions of Third World–controlled production processes. In such situations, a multinational uses cheap local labor to process imported materials, often only components in a larger manufacturing process, into manufactured goods that are sold at a high profit on world markets. A Third World nation then becomes largely a way station in an international assembly line. When a multinational decides for reasons of profit to pick up stakes and move its "way station" elsewhere, often nothing is left behind of this "modernizing" experience but empty buildings, an unemployed work force, and all too frequently, ecological wounds.

A few Third World countries—Taiwan, South Korea, and Brazil—have advanced from being export platforms to becoming integrated, self-generating, industrialized national economies. In fact, their products have developed to such a point that Brazilian steel, Taiwanese textiles, and Korean autos have provoked protectionist demands in First World countries. But these exceptional successes have not been and are unlikely to be repeated on any significant scale in the Third World as a whole. Brazil is unique in the richness of its natural resources, while South Korea and Taiwan (along with Japan) received massive American financial and technical aid and preferential trade privileges—almost the equivalent of an Asian Marshall Plan—during the Korean and Vietnam War periods. Between 1951 and 1965, the United States pumped about $1.5 billion of economic aid into Taiwan, and much more into South Korea—about $6 billion between 1945 and 1978.

In the rest of the Third World, export platforms have yielded few lasting benefits because the semiskilled and unskilled jobs they generate do not provide the technical training needed for local development, because the low wages they pay offer no way out of a domestic market inadequate for local industrial development, and because the multinationals do little more than the former colonial powers to encourage the development of a local industrial base. Multinational plants come and go in an unceasing search for the cheapest possible labor: from New England to the American South, from Mexico to Taiwan, from the Philippines to Indonesia, from Ceylon to Africa, they move their production parts on a never-ending global circuit. This mobility, made possible by high technology, means for most of the Third World a legacy of chronically exploited unskilled labor and abandoned, ecologically degraded industrial parks.

In the final analysis, economic power determines the shape of the world today, and where this power now resides is evident in this matter-of-fact observation by the head of West Germany's largest commercial bank:

> A prime necessity . . . is the improvement of the investment climate in the developing countries themselves, as well as an improvement in the whole attitude towards business activity. . . . In the longer term the necessary investment climate will be created by sheer force of circumstance, because automatically investment capital will flow to those countries providing the necessary conditions—and there are already a number of them. The others will undoubtedly learn the lesson and follow suit in their own interest.[40]

This banker's confidence that sovereign nations will "learn the lesson and follow suit" suggests that nominally independent countries all too often are in reality "company countries" in the same sense that, at an earlier stage of capitalism, Bethlehem, Pennsylvania (Bethlehem Steel), Bisbee, Arizona (Phelps-Dodge Copper), and Butte, Montana (Anaconda Copper), were "company towns." Just as the fate of so many U.S. company towns today is to be sacrificed to multinational interests and to wither away into ghost towns, so the fate of company countries is to fall further and further behind in the global growth Olympics.

The gap in average per capita income between what we now call the First and Third Worlds was roughly 3 to 1 in 1500. Since then it has widened exponentially: 6 to 1 by 1900; 14 to 1 by 1970; and an estimated 30 to 1 by 2000. Sociologist Immanuel Wallerstein concludes that the great majority of the human race at the base of the global economic pyramid is worse off today than in precapitalist times:

> Without any attempt to romanticize the nature of a peasant's life in the Middle Ages, either in Europe or anywhere else in the world, let me offer this brief analysis. If you compare similar strata of the population of the world-economy as a whole, with 70–80% at the bottom, this "bottom" appears worse off today than they were 500–600 years ago, and the top 20% is unevenly spaced geographically throughout the world, and live primarily in such countries as the United States, France, Britain, Germany and Japan. Those who make up the top 20% of the world population may comprise up to 50–70% of the population of those industrial countries. Consequently if someone from one of those countries asks, "Are we better off than our ancestors were 500 years ago?" their answer is not only "yes," but obviously "yes." But that is because those countries have a high percentage of the world's upper strata.[41]

The socioeconomic disruption inflicted on Third World peoples by High-Tech Capitalism has been accompanied by an equally disquieting series of cultural disruptions. Traditional cultures in these countries had been unsettled and traditional elites "Europeanized" or new European elites created by the pre–World War II European empires. Schoolchildren in France's African colonies read textbooks that began with the words "Our ancestors, the Gauls"; while in India, the British official Thomas B. Macaulay declared his intention in 1835 to form "a class of persons, Indian in blood and color, but English in taste, in opinions, in morals, and in intellect." Such a class of deracinated Indians was indeed created, as Jawaharlal Nehru acknowledged: "We developed the mentality of a good country-house servant. . . . The height of our ambition was to become respectable and to be promoted. . . . Greater than the victory of arms or diplomacy was this psychological triumph of the British in India."[42]

Infinitely greater is the psychological triumph of today's high-tech corporate culture. Indian scientist A. K. N. Reddy has noted that technology "is like genetic material. It carries the code of the society in which it was born and sustained, and tries to reproduce that society . . . its structure, its social values."[43]

This is what we are witnessing today, as high-tech culture reproduces throughout the world the "social values" of its Western (and Japanese) home societies. The multinationals do, in fact, determine what is to be manufactured, and then stimulate global consumer demand to absorb the output through familiar advertising techniques. The rise of multinational corporations inevitably led to the globalization of Madison Avenue–style advertising. While in 1954 the top thirty U.S. advertising firms derived only a little over 5 percent of their revenues from overseas accounts, by 1975 the figure had increased to over 50 percent for the largest American advertising agencies like McCann-Erickson, Ted Bates, and Ogilvy & Mather.

This unofficial cultural imperialism of the era of High-Tech Capitalism has proven more influential than the officially sponsored cultural imperialism of the nineteenth-century imperial powers. This is because it touches not merely a tiny Western-educated minority, but the great masses of Third World peoples. If literate, they are reached by billboards, comic books, newspapers, and magazines like *Reader's Digest,* which is published in 101 countries and has a total circulation outside the United States of 11.5 million. If illiterate, they are still reached by TV (and radio) networks, whose showings of "Dynasty," "Dallas," and other U.S.-produced programs have as devoted a following overseas as they do at home.

A study of eighteen hundred primary schoolchildren in Mexico City revealed the extent of television's imprint on young minds: 96 percent recognized TV cartoon characters, only 19 percent the last Aztec emperors; 96 percent identified a local TV character, but only 74 percent could name the then President López Portillo; and more children knew the times of television programs than the dates of religious festivals, including Christmas.[44]

The Third World media blitz has had side effects quite beyond its successful attempts to promote a consumerist mentality. Billboards, magazines, and TV screens tend to depict European men and women as the creators and practitioners of the good life. Such "white is beautiful" advertising inevitably reinforces Third World feelings of inferiority and self-denigration that date back to colonial times. A deliberately fostered consumerist mentality also inhibits local attempts at social change by creating ego involvements that uphold a status quo. Governments trying to cope with basic social and economic problems find it difficult to mobilize support for reducing the immediate supply of familiar consumer goods in favor of future goals such as low-cost housing, mass education, and infrastructure development. Finally, the often mind-numbing programming serves to distract attention from domestic problems, large and small, and is deliberately used for this purpose by authoritarian regimes. A former Philippines information minister has described how President Marcos seized control of all television stations and then embarked on an "idiotization program" consisting of saturation scheduling of foreign-made cartoons, adventure films, soap operas, and situation comedies.[45]

The First World

The destructive effects of High-Tech Capitalism are hardly restricted to the Third World. In the mid 1970s, the great postwar boom in the First World gave way to "stagflation"—a disconcerting combination of stagnation in the economy and inflation in prices. A basic reason for this economic deterioration was the inherent shallowness of the boom, which excluded the great majority of the human race living in the Third World. It was immediately profitable for First World multinationals to move runaway plants to Third World countries in search of labor costing less per day than it did per hour at home. Such low wages, however, put a cap on local purchasing power at a time when millions of peasants were being uprooted from their rural homes, further depressing Third World markets. Hence, the anomaly of the United States selling more goods to Canada with a population of 24 million than to India, Pakistan, and Indonesia with a combined population of 924 million. The basic weakness of the impressive global economy created by High-Tech Capitalism is that it rests on integrated global production but lacks integrated global purchasing power.

This structural defect has only been worsened by a disproportionate drop in the prices of raw materials exported by Third World countries, and by the increasingly heavy external debt of those countries—about 1.2 trillion dollars by 1987—the very servicing of which threatens to push a number of them over the brink into bankruptcy. In order to meet only the interest payments on these debts, Third World governments were often required by their creditors to adopt "austerity" measures, which included reductions in imports. But one-third of all the exports from the industrialized countries go to the Third World, so austerity in the Third World has meant increased unemployment in the First.

Such First World unemployment has also increased because of High-Tech Capitalism itself. Its unceasing technological advances have proved, *in toto,* not merely labor-saving but labor-replacing. Between 1973 and 1985, industrial output in the United States increased almost 40 percent, but the number of blue-collar workers decreased by 5 million. Similarly, in Japan planners anticipate a doubling of industrial production in the next fifteen to twenty years, but a simultaneous 25 to 40 percent cut in blue-collar employment.

The combination of labor-replacing technology at home and inadequate purchasing power abroad is considered responsible for a jump in western Europe's unemployed from an average of 3.4 percent in 1970 to 11 percent in January 1988. In that month, Spain's unemployment rate was 20 percent, Ireland's 19 percent, and Italy's over 14 percent. These statistics lead Peter Drucker to the conclusion that the uncoupling of the raw-material economy from the industrial one is being paralleled by "the uncoupling of manufacturing production from manufacturing employment."[46] The first uncoupling is devastating the underdeveloped countries, the second unsettling the developed ones.

First World unemployment percentages are still far below those of the Great Depression. Yet some workers feel, with justification, that they are in a certain sense worse off now. During the 1930s, unemployment was assumed to be cyclical, and therefore bound to end with the next transition from bust to boom. Today, with jobs being permanently wiped out by robots, computers, runaway industries, and anemic Third World markets, this assumption no longer can be taken for granted—especially since nuclear weapons make a new war-induced boom out of the question.

While factory jobs are decreasing, service jobs like those of bank tellers, fast-food workers, hotel workers, and recreation and health attendants are on the increase. But wages in these service jobs as well as in the high-tech sector of the economy are substantially lower than in the manufacturing one. Whereas American workers in the 1950s and 1960s could look forward to earning one-third more than their fathers, in the 1980s they can expect to earn 15 percent less. Between 1972 and 1987, the average real hourly wages of American workers dropped 10 percent. This amounts for many to the loss of "middle-class" rewards such as home ownership and a college education for one's children. Nor is this phenomenon limited to factory towns. The number of U.S. family farms has dropped steadily from a peak of 6.8 million in the mid-1930s to 2.2 million in 1985. In human terms, this has meant a 40 percent increase between 1979 and 1983 in the number of U.S. rural poor, defined by the Census Bureau as a family of four with a gross annual income in 1983 of less than $10,180. Infant mortality rates in rural U.S. counties were 16.29 per 1,000 live births in 1983, as against 11.3 in urban counties.[47]

The Second World

Socialist countries, too, have experienced the impact of capitalism's creative destruction, albeit less directly than have capitalist and Third World countries. The failure of socialist societies to generate anything comparable to capitalism's technological dynamism has left them unable to insulate themselves from capitalism's inherent drive to integrate the global economy under its aegis. This has proven a striking reversal of the pre–World War II situation in which world capitalism to all appearances was on the defensive and socialist Russia in the ascendant position.

In 1961, in an exuberant moment, General Secretary Nikita Khrushchev boasted that Soviet industrial output would surpass America's by 1980. In fact, the precise opposite happened, with the Soviet economy falling further behind in almost every area imaginable. As early as 1970, the outstanding Soviet physicist and dissident Andrei Sakharov was already warning his compatriots that a "second Industrial Revolution" was under way in the West, one with which the Soviet Union was "incapable of keeping abreast." The basic responsibility for this failure, he argued, was not the socialist system per se but the lack of democracy in the Russian version of that system. Without democracy, without the free exchange of ideas in the scientific community and in Soviet society as a whole, fear would lead to a conformity and rigidity that would paralyze the entire Soviet system. "From our friends we sometimes hear the U.S.S.R. compared to a huge truck, whose driver pressed one foot hard down on the accelerator and the other on the brake. The time has come to make more intelligent use of the brake!"[48]

Not only was Sakharov's prophetic warning ignored, but he himself was vilified and persecuted during General Secretary Leonid

Brezhnev's long stewardship (1964–82), while the brake was slammed down even harder. At a time when computers were permeating and regenerating the Western and Japanese economies and societies, Soviet authorities viewed even copying machines as potential instruments of subversion and kept them under lock and key, strictly regulating and recording all access to them.

Such self-defeat by ham-fisted control from above was ludicrously evident in Stalin's sponsorship of a pseudoscience called "agrobiology," propounded by Trofim D. Lysenko, which led to thirty-five years (1929–64) of disasters, agricultural and scientific. Just as government theoreticians maintained that human nature could be changed by altering the economic and political structure of society, so Lysenko claimed that heritable changes could be made in plants and animals by altering their environment with such measures as grafts, new feed procedures, the selection of individual seeds, and the special treatment of seeds before germination. Geneticists overwhelmingly rejected Lysenko's "science," but government officials welcomed and supported it, paralleling as it did their ideological preconceptions, and promising to boost quickly and inexpensively the productivity of the lagging collective farms. So a vigorous Russian school of genetics was destroyed, while Lysenko was made president of the Lenin Agricultural Academy. The end result was a protracted and embarrassing aberration in the history of Russian science, with disastrous repercussions for Soviet collectives as well as Soviet laboratories, and the ceding of the basic work in genetic engineering to world capitalism.

Under such regimentation, high-tech experimentation and invention never took off. Instead, a scientific and economic ossification set in, as reflected in a sharp decline in the growth of the Soviet GNP during the 1970s. Russia not only failed to surpass America, as Khrushchev expected, but instead fell behind Japan, and now ranks number three in the global industrial hierarchy.

General Secretary Mikhail Gorbachev has candidly revealed the extent of this Soviet debacle. Economic growth had slowed down so seriously that by the beginning of the 1980s, according to Gorbachev, it "had fallen to a level close to economic stagnation." The full impact of this failure was not felt at that time because the Soviet Union's supply of human and material resources was so plentiful. What was lacking technologically was compensated for by the lavish use of those still easily available resources. The twelvefold increase during the 1970s in the global price of oil and gas was a bonanza for the Brezhnev regime, since these were the Soviet Union's principal exports. The swollen oil and gas revenues made it possible to buy the approval of most segments of Soviet society. Working-class families for the first time enjoyed refrigerators, television sets, and marginally improved housing; officialdom happily filled the new positions afforded by an expanding government and party bureaucracy; while the military received sufficient funds to gain strategic parity with the United States. This explains why the Brezhnev era, though widely perceived as one of stagnation, cronyism, and corruption, was also one of complacency and stability.

The precarious foundation of Brezhnevian prosperity has been analyzed by Soviet economist Stanislav Menshikov, former U.N. official and staff member of the Communist Central Committee. He has noted that the Soviet government was able to draw on its rich oil reserves to export enough to pay for vast wheat imports from the West. These were used mostly as feed grain for livestock, enabling the Soviet public to eat more meat and less bread. Thus, finite oil reserves were being depleted to offset the inefficiency of Soviet agriculture, rather than for buying the sorts of Western and Japanese machinery and technology that might have helped modernize both Soviet industry and agriculture. In this respect, there is an instructive parallel between the Soviet Union today and Spain in the sixteenth and seventeenth centuries, a time when a flood of New World bullion buttressed an obsolescent Spanish economy. Semifeudal Spain already was an economic dependency of capitalistic northwestern Europe when the treasure galleons began arriving with their cargoes of colonial booty. The booty served to fossilize and shore up the antiquated Spanish society of early modern times, just as Menshikov now fears that oil, gold, and other Soviet natural resources are having a similar fossilizing effect on a similarly obsolete Soviet society.[49]

Not only Menshikov but Gorbachev himself has explicitly warned that the Soviet economy is "still spending far more on raw materials, energy, and other resources per unit of output than other developed nations." The country's natural wealth, concludes Gorbachev, "has spoilt, one may even say corrupted us." That wealth enabled Soviet planners to cope superficially with shortages and other emergencies as they arose, thus evading in the short run the basic structural problems that were forcing the country further and further behind the capitalist world economy.

Such a sweep-it-under-the-rug approach to economic problems has become a more and more dangerous luxury for Russian planners. Even seemingly inexhaustible Soviet resources like oil and ores are showing the first signs of depletion. Soviet planners face another problem as well. National revenues are shrinking because of a sharp drop in the world price of oil, the Soviet Union's chief export, and an even sharper drop in key vodka tax revenues because of Gorbachev's drive against alcohol consumption—part of his overall campaign to create a more productive, cost-effective modern work force. It is against this backdrop that Gorbachev's comprehensive program to reform and modernize the Russian economy must be viewed.

To a remarkable degree, Gorbachev makes the same diagnosis and prescribes the same remedy as Sakharov did two decades earlier. Even the images he employs bear a certain similarity to those of the physicist. While Sakharov, for instance, talked of brakes and accelerator being pressed at the same time, Gorbachev uses the image of the huge flywheel of a powerful machine that continues to revolve, but with its drive belts too loose. Given this basic harmony in the views of the two men, it is not surprising that, by the spring of 1988, Sakharov had been released from internal exile and was stating at a Moscow news conference: "Mikhail Gorbachev is an outstanding statesman . . . one of the big doers in *perestroika*. From the bottom of my heart, I wish success for the cause with which his name is associated."[50]

Gorbachev's remedy, like Sakharov's, is not the abandonment but the purification of socialism from "the cult of personality, the system of management by command . . . and bureaucratic and dogmatic aberrations . . . that have led to stagnation. These phenomena . . . should become things of the past." Gorbachev's *perestroika,* or restructuring, is designed specifically to correct these "distortions." "*Perestroika* means mass initiative," states Gorbachev. "It is the comprehensive development of democracy, socialist self-government, encouragement of initiative and creative endeavor. . . . It is utmost respect for the individual and consideration for personal dignity."[51]

Implicit in Gorbachev's analysis is the assumption that once current "aberrations" are corrected, socialism then will be safely back on track and will demonstrate its virtues. Despite his strenuous efforts, Gorbachev's program has encountered many obstacles and seems to have made little real progress in reconstructing Russian society. Three-and-a-half years after he began, he acknowledged in a candid speech before Soviet editors that *perestroika* remained mostly rhetoric, mostly a distant goal. "We are going slowly, we are losing time, and this means we are losing the game. In a word, it turns out there is a gap between our goals and our work. . . . it is important to emphasize in no uncertain terms that no radical breakthrough has occurred and that the economy has not emerged from a state of stagnation."[52]

Since "stagnation" and failure to achieve a "radical breakthrough" are the common experience throughout the socialist Second World, it is apparent that the impediment is not exclusively or peculiarly a Soviet phenomenon. It is broadly systemic in nature rather than narrowly national. The afflictions and the prospects of contemporary socialism will be analyzed in the concluding chapter.

THE NEW TIME OF TROUBLES, 1975 AND AFTER

As the twenty-first century approaches, the world is once again experiencing a new time of troubles. The troubles this time are, however, of an entirely different order of magnitude—more comprehensive and more basic—reflecting the qualitative difference between Industrial and High-Tech Capitalism. They are more comprehensive because they are not confined to any specific region or type of society. The socialist Second World and underdeveloped Third World are in even more serious trouble than the capitalist First World. The troubles are more basic because they have transcended specific regional difficulties to threaten planetary well-being, and even, in the case of "nuclear winter," the existence of the human species.

Historian Arnold Toynbee, who had been aware of the onset of the first time of troubles in the 1930s, sensed in the 1970s the onset of a new time of troubles, whose ecological and nuclear perils he articulated. "It is possible that people who are already alive might have their lives cut short by a man-made catastrophe that would wreck the biosphere and would destroy mankind together with all other forms of life."[53]

Today's troubles are especially intractable, affording as they do no way out comparable to the one provided by the preparations for World War II, which revived economies around the world and finally ended the Great Depression. A World War III would mean not an answer to the problems of the global economy, but an irreversible way out of the world itself.

LIFELINES

ECOLOGY

On April 9, 1983, Paul J. Weitz, commander of the first flight of the space shuttle *Challenger,* was returning to home base on planet Earth. As he neared his destination, he was struck by the deterioration of what originally had been a beautiful blue planet. "It was appalling to me to see how dirty our atmosphere is getting. Unfortunately, this world is rapidly becoming a gray planet. Our environment apparently is flat going downhill. . . . What's the message? We are fouling our nest."[54]

Indeed, we are fouling the thin and fragile layer of atmosphere, soil, and water that comprises our "nest." As noted earlier, ecological stresses are not peculiar to our age. Early hunter-gatherers left at least a faint mark on their environment when they lit their campfires and set woods ablaze to drive out game. From China to Peru, later agricultural peoples made a far more significant imprint when they cleared out entire forests and constructed vast irrigation systems. Today, however, the ecological impact of humans on their planet has become explosive because of three major interrelated factors: the exponential growth of human populations, the imperatives of the new global capitalist economy, and the hardly foreseen consequences of a seemingly uncontrollable technology.

From the seventeenth century on, Europe led the way in population growth as its increasingly productive agriculture and industry provided the means to sustain a rising birth rate. Later, advances in medical science and the adoption of numerous public health measures reduced death rates as well. Consequently, Europe's population soared from 100 million in 1650 to 463 million in 1914. Since Europeans at the same time were also colonizing overseas territories, the percentage of people of European origin rose from less than one-fifth to about one-third of the total world population in less than three centuries.

In recent decades, this particular global demographic trend has reversed itself. Birth rates in the developed regions have plummeted so that they match the low death rates, and a virtual population equilibrium now prevails in Europe, North America, Oceania, and Japan. In the underdeveloped Third World, by contrast, recently increased food production and improved health technology have sharply lowered death rates while birth rates soar. Third World population growth is presently twice that of the developed world—2.2 percent as against 1.1 percent—a demographic pattern likely to persist regardless of current population-control campaigns because Third World populations are predominantly young. The prospects are a 50 percent increase in world population between 1980 and 2000, and over a 100 percent increase by 2025.

POPULATION PROJECTIONS: 1950–2100
(Population in Millions)

	1950	1980	2000	2025	2050	2100
Underdeveloped Countries	1,670	3,284	4,922	7,061	8,548	9,741
Developed Countries	834	1,140	1,284	1,393	1,425	1,454
Total World	2,054	4,424	6,206	8,454	9,973	11,195

Source: U.N. and World Bank estimates and projections.

These statistics are unprecedented in their magnitude. The human race reached its first billion in 1 million years, its second billion in 120 years, its third billion in 32 years, its fourth in 15 years, and its fifth in 10 years. These statistics also represent an unprecedented situation because the ecological setting for such demographic growth has changed radically. In the past, relatively slow population increases were always at least partially accommodated by the availability of "empty" lands. In ancient times, the Indians migrated southeast into the Ganges valley, and the Chinese south into the Yangtze valley and beyond. Western Europeans of medieval times migrated eastward in large numbers into the underpopulated lands of central and eastern Europe. More spectacular yet was the tidal wave of Europeans who flooded into the "empty" lands of the Americas and Oceania. Today, however, when world population increase far outstrips any in the past, no more "empty" territories remain. The result can only mean rising social tensions, with the number of landless rural households in India alone increasing from 15 million in 1961 to an estimated 44 million by the year 2000.

World population increase engenders a corresponding environmental stress that is compounded by the unprecedented disruptiveness of modern technology. Even in the United States, which is relatively underpopulated and richly endowed, evidence of such environmental stress is everywhere—in the polluted, water-scarce sprawl of cities like Los Angeles and Phoenix; in the depletion of the great Ogallala Aquifer, which supplies irrigation water to the Great Plains from Nebraska to the Texas Panhandle; in the conversion of California's orchards and Long Island's potato fields into housing developments; in the clear-cutting of remnant primeval forests in the Northwest; in the regions where half of all Americans live where air pollution exceeds health standards; and in the 99 percent of known toxic dumps that have not been cleaned up and that are contaminating water supplies throughout the country.

The global capitalist economy has left an even deeper imprint of environmental stress on planet Earth, one of whose early manifestations was the cutting down of New World forests to make way for sugar, cotton, rice, and banana plantations. Even today Haiti and northeastern Brazil are the most deforested and the poorest regions in the Western hemisphere, a linkage that is less than coincidental. Such a pattern was repeated centuries later in West Africa for the sake of cotton and peanuts, and in Southeast Asia for rice, tea, coffee, and pineapples.

Precisely the same imperative is still operative today, and with the same consequences. In the early 1970s, when world demand for wheat was high and the supply low, Agriculture Secretary Earl Butz exhorted American farmers to plant their land "from fence row to fence row" to maximize production. The farmers responded enthusiastically, tearing out rows of trees that had been planted as windbreaks after the 1930s Dust Bowl. As a result, American farmers once more are seeing their topsoil being blown away by the wind.

Similarly, the First World demand for tropical hardwood furniture and wall paneling has tempted cash-poor governments in Africa and Southeast Asia to allow their forests to be overexploited. According to U.N. estimates, Africa has lost 23 percent of its forests since 1950, and the Himalayan watershed 40 percent—a depletion that is largely responsible for the devastating floods that have in recent years ravaged Nepal and Bangladesh. One key cause for the speed-up in global deforestation is cattle ranching in tropical forests, primarily to produce beef for American fast-food outlets. Once the trees are cleared away and the relatively nutrient-poor tropical soils planted with grass, the consequences are devastating. Erosion and leaching by driving rains, and oxidation and decomposition by the sun's heat, quickly lead to massive soil degradation. Pasture grasses are displaced by weeds, many of which are toxic to cattle. About 15 percent of cattle in the Amazon die each year from eating such poisonous weeds. After five to seven years, the productivity of these new grazing lands usually falls dramatically and the ranchers move on to clear new rain-forestlands. In that short time, however, damage has been inflicted on the earth's tropical forests that is much more difficult to reverse than it would be in temperate zones. Since 1960, it is estimated that three-fourths of all Central American forests have been destroyed for the sake of lowering the price of hamburgers in the United States by a nickel. In addition to these external commercial pressures, deforestation all over the globe continues because of the desperate efforts by landless peasants to carve family plots out of forestland for their subsistence agriculture.

The ecological repercussions of the accelerating pace of technological development worldwide are symbolized by a succession of catastrophes that remain fresh in collective memory. Some have been dramatic and attention-grabbing, as the names Hiroshima, Bhopal, Chernobyl, and Prince William Sound indicate. Others, while less obvious, have perhaps been even more fundamental, reflecting as they do the possible crossing of certain key thresholds of danger in the balance of planetary ecosystems. Such "crossings" may include the recently discovered "hole" in the earth's shield of ozone, which has protected earthly life from the sun's ultraviolet light. Not until the ozone layer formed, 500 million years ago, could early life forms creep out onto the land. Until then, they had to seek shelter in the sea—an indication of the magnitude of the possible repercussions if the depletion of the ozone layer continues un- checked.

Another impending "crossing" now goes by the name "green-house effect." This effect, scientists speculate, could be produced by certain waste gases that let in sunlight but trap heat that otherwise would escape into space. One of these, carbon dioxide, has been steadily building up in our atmosphere through the burning of coal and oil, and because forests that once absorbed excess carbon dioxide are rapidly being destroyed. Although scientists disagree as to the validity of the "greenhouse" theory, it has been noted that five of the last nine years—1980, 1981, 1983, 1987, and 1988—have been the warmest since measurements of global surface temperatures began a century ago. It is also acknowledged that even a slight warming of global temperatures could trigger drastic climatic changes. The Gulf Stream might shift course and fail to warm Europe, which would thereafter experience the frigid Arctic temper-atures of Labrador, which is in the same latitude. Ocean levels might also rise twenty feet if the Arctic and Antarctic ice caps melted, flooding entire countries such as Holland and Bangladesh, as well as coastal cities from New York to New Orleans.

As already discussed, another "crossing"—an actuality rather than a hovering peril—lies in the proliferation of deforestation worldwide. Since the dawn of agriculture some ten thousand years ago, the earth's forests have shrunk from 6.2 billion to 4.2 billion hectares—a loss of one-third. For centuries the clearing of forests was welcomed as the prerequisite to raising food production and general economic and social development. Today, with forested areas drastically reduced and the world's population even more drastically increased, the continuing deforestation is responsible for serious ecological damage, including erosion, desertification, flood-ing, and reduced land productivity, as well as fuel loss for the hundreds of millions of Third World peoples who gather wood to cook their meals and heat their homes. Rain forests are also the source of countless thousands of useful plants, animals, and insects, many of which have not yet been discovered, much less investigated in terms of their value to humanity. In this light it is worth noting that one-quarter of all prescription drugs in the United States are derived from rain-forest plants. A serious side effect of deforestation then may be the shrinkage of gene pools available to scientists and farmers. Finally, forests play such a crucial role in the global cycling of carbon dioxide that tropical forests, especially those of the Amazon valley, have been characterized as the earth's "lungs." Continued decimation of those forests appears to be building up to a condition of global emphysema.

Summarizing the overall significance of these ecological trends, the Worldwatch Institute warns: "The threats that emerge as we cross natural thresholds are no longer hypothetical. . . . No generation has ever faced such a complex set of issues requiring immediate attention. Preceding generations have always been concerned about the future, but we are the first to be faced with decisions that will determine whether the earth our children inherit will be inhabitable."[55]

A Soviet journal stated flatly in 1979 that "capitalism is pushing society towards an ecological catastrophe. . . . True harmony between nature and society can be achieved only under conditions of socialism with its integral, humane social relations."[56] Unfortunately, the gathering ecological crisis is more complex and obdurate than is suggested by this analysis. In actual practice, socialist countries have proven at least as ecologically destructive as the capitalist ones, if not worse. They do not have to contend with corporations for whom the "bottom line" is the dividing line; but they do have to contend with institutional impediments that are as serious as any under capitalism.

A major impediment for centrally planned economies seeking to improve the environment is the inefficiency inherent in their current mode of operating. Being until recently free of market-pricing pressures, they lack the incentive to reduce the quantity of inputs used in the production process. The result is much higher resource consumption per unit of GNP in socialist countries than in capitalist ones, which means correspondingly higher pollution levels. Energy expenditure (in megajoules) per dollar of GNP in 1983 was 8.6 in France and Sweden, 9.7 in Japan, 11.8 in West Germany, and 19.3

in the United States, as against 21.5 in Yugoslavia, 32.3 in the USSR, 40.9 in China, and 49.5 in Hungary.

Another impediment for centrally planned economies is that the penalties for violating environmental regulations are usually far lighter than those for not meeting production goals, and so have little if any deterrent effect, especially since the penalities are usually paid by the government itself. As one Soviet observer aptly commented; "Forcing enterprises without economic incentives to control pollution is like getting a cat to eat cucumbers by giving it a lecture on the benefits of vegetarianism."[57]

Given such institutional obstacles, it is not surprising that environmental degradation is as acute in eastern Europe as anywhere else. The Polish government has declared five villages in the industrial region of Silesia unfit to live in because of heavy metal contamination of the soil. In Czechoslovakia, 4,300 miles of river—28 percent of the nation's total—has no fish life, and 70 percent of the rivers are considered heavily polluted. "When trees are dying," observes Czech activist Lena Mareckova, "when the water is dirty, when the air is dangerous for little children, that's worse than prison."[58]

An American scholar, after studying the Soviet ecological situation, expressed his "déjà vu" impression in this fashion. *"Homo economicus,* whether socialist or capitalist, seems far more concerned with maximizing production and profit than with minimizing environmental damage. And the same factor in each system—that it is cheaper to continue than to abate pollution—impels *Homo economicus* to behave as he does."[59]

Whatever the social system, the central issue is one of priorities. It is an issue that was clearly perceived by Francis Bacon, that giant of the seventeenth-century Scientific Revolution. He enthusiastically endorsed the pursuit of "knowledge and skill" through science, but he added that the pursuit should be conducted with "humility and charity," and *not* "for pleasure of mind, or for contention, or for superiority to others, or for profit, or fame, or power, or any of these inferior things; but for the benefit and use of life."[60] Bacon's choice of humility, charity, and life over profit, fame, and power merits reflection at a time when the achievements and perils of his beloved science surpass anything that even his genius could have imagined.

GENDER RELATIONS

The dilemma of values or priorities is as relevant to gender relations as to ecological relations. Technology has again been a decisive catalyst in the nature of relations between the sexes, whether in the form of eighteenth-century textile machines that gave women employment opportunities outside their homes, or today's birth-control pills that give women more control over their bodies and careers. The net result has been to begin to reverse the effects of the Agricultural Revolution and of the ensuing tributary civilizations that ended the gender equality of kinship societies and confined women to work in or around the home. Successive technological advances have enabled women to break the bonds of endless pregnancies and of housework, and to enter the outside work force in numbers comparable to those of men. Customs and values have, however, failed to keep pace.

The timing, course, and impact of the Industrial Revolution varied fundamentally from region to region, and its impact on gender relations varied correspondingly. Meaningful analysis of these relations therefore requires separate treatment of the evolution of gender relations, first in the developed and then the underdeveloped worlds.

The Developed World

The immediate effect of the eighteenth-century Industrial Revolution on women was to pressure or entice them out of the family economy and into the new wage economy outside the household. In certain respects, this represented a step forward. The preindustrial family economy is commonly associated with pleasant and wholesome family activity. In reality, for women it involved much repetitious and monotonous work, such as spinning, tatting, and the setting up of looms. Women often worked long hours beside their fathers or husbands, yet still faced the onerous tasks of housekeeping and child care. With the shifting of the primary workplace from home to factory, women, who were considered a cheaper and more docile work force than men, found steady employment because industrialists, who had much capital invested in their plants, closed down operations as little as possible. Many earned incomes of their own for the first time or received substantially higher incomes from factory work than they had previously received from piece work in family quarters. Women in the Manchester cotton factories in 1914, for example, earned more than double the amount paid to those who hemmed handkerchiefs at home.

The new factory jobs, on the other hand, had certain negative features detested by both male and female workers. These included long inflexible hours, unhealthy working conditions, and myriad regulations and penalties. These were especially galling for peasants who had adapted their lives to the rhythm of seasons and of sunrise and sunset, but who as factory workers had to submit on pain of fines or dismissal to bells that tolled the hours and minutes, and to foremen empowered to enforce the regulations as the unchallengeable bosses of the workplace.

Women were especially vulnerable to exploitation because they were less likely to be organized into unions to defend themselves; male union officials usually did little to encourage female membership. Also, women, still saddled with children and housework, had less free time left for participation in union activities. It is hardly surprising, then, that the average wage for female factory workers at the end of the nineteenth century was less than half that of their male counterparts.

The move out of the house was for some women a move not only into new economic but into new political activities—though women were ordinarily left to serve only as auxiliaries in male-controlled political movements. Their support was naturally welcomed during any struggle for power—often with grandiose promises of female liberation from social oppression to follow—but after victory was won, they were at best ignored and often forced to return to something approximating their original subordinate status. This pattern has prevailed in all modern revolutions, from the English in the seventeenth century to the Russian and Chinese in our own.

During the French Revolution, for example, middle-class women set forth their demands in their *cahiers* ("notebooks") to the Estates General, while working-class women took direct action by storming the Versailles Palace when there was no bread for their families. When revolutionary France was invaded by counterrevolutionary armies, women joined in the defense of the young republic by rolling bandages in hospitals, making shirts and trousers for the soldiers, and even fighting in the revolutionary armies. Grateful republican governments responded with legislation legalizing divorce, granting wives a share of family property, and providing free and compulsory primary education for girls as well as boys. But when the foreign danger passed, a general reaction set in during which the newly won women's rights were scrapped. Napoleon capped this reaction by including provisions in his new legal code that restored the absolute authority of father and husband.

One reason for this setback was that the cause of women's rights did not have priority among the women themselves. During the revolution, they had responded primarily to the needs of their class rather than of their sex. They had joined the struggle mainly for social and economic relief rather than for female rights. So when the French Revolution turned conservative, as had the English Revolution earlier, and as would other revolutions to come, the gains of the women as females were swept away along with their gains as workers. Political leaders then resurrected familiar speeches about the future contributions that women could make within the family, not outside.

Similar disappointment awaited the women of the developed countries after they won the right to vote during the course of the twentieth century. Many had assumed that they would gain their goals if they shared political power with men on an equal basis. Therefore, the right to vote became the chief demand of suffragettes, and gradually they won that right. The number of countries with universal franchise increased from 1 in 1900 to 3 in 1910, 21 in 1930, 69 in 1950, and 129 in 1975. But enfranchisement did not automatically confer political power. Few women were elected to representative bodies, and fewer still ended up in positions of executive authority.

In the second half of the twentieth century, new horizons have opened up for women in all developed countries, both capitalist and socialist. A variety of contraceptives have given women control over their own reproductive functions and the possibility of greater equality in the realm of sexuality, allowing them, for instance, to have sex while refusing motherhood, or to space pregnancies to allow for the continuation of careers. Something approaching equality of opportunity in education has also developed in the First World, where male and female students now are attending school in roughly equal numbers at all educational levels. A further broadening of horizons has occurred in the work force, where the two sexes now are approaching numerically equal participation. In the Soviet Union, the percentage of women in the work force has risen from 24 in 1928 to 50 in 1984. In the United States, the percentage of adult women working outside the home has risen from 18.9 in 1890 to 25.8 in 1940 to 54 in 1984, when women made up 44 percent of the total American labor force. By 1986, in fact, American women had surpassed men (6.938 to 6.909 million) in the number of "professional" jobs they held—the category including architects; engineers; mathematical, computer, and natural scientists; physicians; dentists; pharmacists; teachers; librarians; social workers; and lawyers.

These historic advances have not been an unmixed blessing. On the positive side, women now are less dependent and enjoy new opportunities for realizing their potentialities. There is also the economic advantage of two paychecks, which enables many families to enjoy larger homes, longer vacations, and more education for their children. On the other hand, having taken on full-time jobs, women now find themselves overloaded, because household chores and child care continue to be viewed as women's work. American women today work an average of 24.2 hours per week at home, compared to their husbands' 12.6 hours. In the Soviet Union, the disparity is even greater: 25 to 28 hours per week by Soviet women, 4 to 6 hours for the men.

The hold of tradition is clear also in the all-pervasive wage discrimination women face because of the time-honored assumption that work performed by women is, by definition, less important and less valuable than that performed by men. We read in the Bible (Lev. 27:3–4) that the Lord said to Moses: "Your valuation of a male . . . shall be 50 shekels of silver. . . . if the person is a female, your valuation shall be 30 shekels." In the two thousand years since this biblical injunction, the appraisal of the financial worth of the two sexes seems to have changed little. In both the Soviet Union and the United States, the median income of full-time women workers today ranges from 60 to 70 percent of male median income.

One reason for this wage differential is that women tend to enter the work force later and with less training because their school education is less job-oriented. Some forego overtime work and promotions that might interfere with domestic obligations. More basic is the unspoken assumption that "women's work" is less valuable, even if it requires more training and skill. "A mistake that a nurse makes kills you" notes Minnesota's employee relations commissioner, "a mistake that a wall painter makes offends you. So why should a painter make more money."[61] The answer, of course, is that most wall painters are male, which is why Denver tree trimmers are paid more than nurses in the city's intensive care unit, and why a liquor clerk in Montgomery County, Maryland, with a high-school diploma earns more than a schoolteacher with a college degree. This disparity has sparked the current demand for "comparative worth"—a shorthand phrase for the proposition that women should get the same pay as men, not only when doing the same jobs as men, but also when doing jobs roughly equivalent in terms of education, skills, and working conditions.

Despite the progress the comparative-worth principle represents, the actual position of women in the United Sates has been deteriorating disturbingly during the 1980s with the end of the "fair-weather feminism" of the affluent post–World War II years. During that era, it was assumed that poverty and cyclical economic crises were on the way out. Feminists therefore focused more on affirmative action, the Equal Rights Amendment, sexist images in the media, and the creation of women's studies departments in colleges and universities. In fact, the number of courses about women offered in American universities increased from seventeen in 1969 to thirty thousand in 1982. The cultural and sexual situations of women—and the development of a new feminist consciousness—tended to be stressed over more strictly economic issues.

The ending of the postwar boom during the 1970s also cast shadows on this "fair-weather feminism," which unsurprisingly did not foresee the future better than the rest of society. Many women now find it necessary to accept inferior job arrangements such as part-time work or home work at lower rates and with no fringe benefits. At the same time, 50 percent of American marriages are ending in divorce, following which the living standard of the average ex-husband rises 42 percent, while that of his wife and children falls 73 percent. This trend means that the two-paycheck nuclear family can no longer be considered the norm. A steadily increasing number of women now are unmarried or single-parent heads of families. In 1986 the median income of male heads of households was $23,000 compared to $13,500 for female heads of households. The plight of

American women is worsened by the lack of the sort of government financial supports and social services available in western European democracies. Hence, the conclusion of sociologist Diane Pearce that "fair-weather feminism" has given way in the United States of the 1980s to the "feminization of poverty."

The Underdeveloped World

If the position of women in the developed world is precarious, that of their sisters in the underdeveloped Third World sometimes seems nothing short of hopeless. When their countries became vulnerable to exploitation by the developing capitalist order, they found themselves the most vulnerable and exploited of all. Worst off in the period of Commercial Capitalism were the female slaves on plantations. Whether they were allowed to "breed" more slaves was decided strictly by cost-benefit calculations. Before the 1760s in the Caribbean, most plantations were small and had few slaves, so the owners encouraged natural propagation of the slave population. After the 1760s, as the plantations became large operations and the slave trade became a vast commercial enterprise, slaves became available in greater numbers at lower cost. Planters then began to calculate that it was "cheaper to purchase than to breed," since slave mothers worked less efficiently before and after a birth, and slave children had to be attended to and fed for many years before a profit could be realized from their labor. So slave pregnancies came to be actively discouraged on the British, French, and Dutch Caribbean sugar islands by the ill-treatment of mothers before and after delivery. When, however, the antislavery movement spread throughout Europe, reducing the supply of slaves worldwide and raising their cost, the Caribbean planters reversed their policy and once again encouraged breeding by local slave women.

In Africa and Asia, the first European traders arrived with guns but not wives, and often married local women. As the number of children from these marriages increased, European authorities became concerned about possible "complications" in the future. European women were then encouraged to emigrate to the colonies, and the status of local women who consorted with European men sank to that of prostitutes or concubines. The British statesman John Strachey described how two of his ancestors serving in India in the eighteenth century married "a Bengali lady of a distinguished family and a Persian princess. . . . without exciting the least adverse comment or injuring their careers in any way." By the late nineteenth century, adds Strachey, "how unthinkable such alliances would have been." Strachey concludes that "this terrible withdrawal of genuine human community undid "the relations of the two great peoples."[62]

The combination of sexism and racism was a powerful fuel that helped to drive the imperial machine of nineteenth-century Europe. It allowed Europeans to degrade their colonial subjects, conveniently seeing them as somewhat less than fully human. Evidence that this racist-sexist combination still characterizes First World/ Third World relations is not hard to find. At its most obvious, it can be seen in government-sponsored "sex-tourism," conducted with particular openness in Southeast Asia.

This phenomenon first took its "modern" shape during the years of the U.S. military presence in Vietnam. Hundreds of thousands of prostitutes converged on "Rest and Recreation" centers for GIs in Saigon, Bangkok, and Manila. In the wake of the withdrawal of American forces from Southeast Asia in the early 1970s, a flood of male "tourists" from Japan, Australia, western Europe, and the United States took up the same R&R circuit. According to Thai police, there were 700,000 prostitutes in their country in 1982—10 percent of all Thai women between the ages of fifteen and thirty. In the same year, the International Labor Organization reported 200,-000 prostitutes in the Philippines and 260,000 in South Korea. Far from being alarmed by these figures, leaders in some of the affected countries have actively promoted what they view as a growth industry, one attracting sorely needed foreign currency. In South Korea, for instance, the Ministry of Tourism issues permits to *kisaengs* after they pass a training course in which they are given anti-Communist indoctrination and also taught the sexual positions "preferred by" and "most appropriate for" Japanese tourists aged forty to fifty. The *kisaengs* are assured by their instructors that "you do not prostitute either yourself or the nation, but express your heroic patriotism."[63]

Many Third World women are also exploited as cheap immigrant labor in First World countries. Those workers send their earnings home, where they contribute crucially to the well-being of their families. Of the 400,000 Filipinos who work overseas, 175,000 are women, who are preferred because, in the words of a London employment agency, "They are more obedient, more like servants. They will do work that, say, Moroccans, Portuguese, and Italians would refuse." The Philippine secretary of labor, Franklin Drilon, has stated candidly: "The Philippines has virtually become a country of maids, cheap domestic labor to clean up after the rest of the world. The situation has not only adversely affected the morale of Filipino women but the country's image as well." That image is being further tarnished by the widespread advertising of Filipino women as "mail-order brides" for lonely men in Asia, Europe, and Australia.[64]

Third World women have also experienced systematic gender discrimination in the workplace in their own countries because of European preconceptions concerning female labor. "Farm work is man's work" was the common assumption of the European overlords, despite the fact that women, especially in Africa, traditionally had done most such work. Colonial officials nevertheless clung to their belief that local women were "breeders and feeders," and therefore excluded them from allocations of agricultural aid whether in the form of loans or of instruction in improved techniques. Consequently, men were trained to produce cash crops, while women continued producing subsistence food crops with traditional techniques. This discrimination has been passed on from colonial times to the present and affects the policies of international organizations today, as it did colonial policies in the past. It is especially evident now regarding ownership of land. Traditionally, husbands were granted full ownership rights, even though their wives might have done most of the cultivating. This has hardly changed today, even when men all over Africa are leaving rural areas to work in cities. There they frequently remarry, leaving their first wives, the very ones actually working the land, legally landless.

Approximately 4 million Third World women have themselves left the land to find employment in factories built by multinational corporations. Enticed by cheap labor, tax exemptions, and legal bans on labor organizing, the multinationals have organized a "global assembly line" stretching from Latin America to Southeast Asia. Although wages are a small fraction of those prevailing in the West, nevertheless they are above local wage levels available to such women and therefore are welcomed. Such jobs also offer young women wider social contacts and a possible alternative to arranged marriages. Local officials, welcoming the transplanted industries as a source of jobs and revenues, urge young females to be "good workers" for the sake of "national prosperity."

Whatever its immediate advantages, this "global assembly line" is not a long-term or comprehensive panacea for Third World women. The new jobs, invariably unskilled and of little value for genuine national economic development, are inherently ephemeral as the multinationals are ever ready to pull up and move to new reservoirs of still cheaper labor. A noteworthy recent development is the extension of the "global assembly line" from Third World countries to the United States itself. The General Accounting Office reports a proliferation of sweatshops in American cities, especially in the garment and restaurant industries. These shops employ Asian and Hispanic female immigrants, frequently at less than the legal minimum wage. The women accept the substandard wages and working conditions because they lack the occupational and language skills needed for better-paying work, and because many are illegal immigrants and therefore afraid to attract any attention.[65]

Another recent development is the appearance of the "global office" alongside the "global assembly line." Improvements in telecommunications and computer technology permit office jobs to be performed thousands of miles from the central office, wherever wage rates are lower. American insurance companies, for example, are opening offices in Ireland, where unemployment rates are three times higher than in the United States, and wages 50 percent lower. Medical claims are forwarded by overnight mail to Ireland, where young women enter the information into personal computers linked by transatlantic line to the data processing centers in the United States. The Irish workers determine the claims and amounts, and transmit the decisions instantly to the United States. Despite the telecommunications and postage charges, the insurance companies are reducing their expenses by 20 percent with the shift to a global rather than a national scale of operations. Likewise, airline companies are saving money by sending ticket stubs to Caribbean offices, where keyboard operators type the flight information into a computer, instead of having more costly workers do that at data-processing centers in the United States.[66]

Women of the developed and underdeveloped lands have had

one common historical experience: both have participated in revolutionary movements, and after the revolutions both have endured regression to the traditional inequality characterizing all tributary civilizations. This pattern, noted in the English and French revolutions, was repeated in the great twentieth-century upheavals in the Third World. Typically, a Vietnamese Communist Party statement issued in the midst of the struggle against American intervention stated explicitly that "the illusion of sexual equality" is a "bourgeois theory," and that women can "emancipate themselves" only through "serious political education" that will "raise their political consciousness and make them participate in working class organizations."[67] This did not dampen the ardor of Vietnamese women, who, during the course of the resistance, made up 80 percent of the rural and 48 percent of the industrial labor force, dominated the revolutionary educational and health programs, and frequently fought with the guerrilla forces. But now that the armed struggle is over, few women are to be found in top policy-making posts in Vietnam (or China or the Soviet Union). They continue to be excluded from leadership posts now, as they had been during the revolutionary years, and earlier. The vice-president of a Vietnamese women's union explains poignantly: "The heritage of Confucianism, feudalism and capitalism runs deep. . . . we still do not have full equality. . . . we cannot liberate women by the stroke of a pen. . . . it is much harder to fight against obsolete customs than against the enemy."[68]

How much harder it is to fight against the hold of the past was experienced anew by women engaged in the Nicaraguan revolution. Unlike many of its counterparts elsewhere, the revolutionary Sandinista Front did not denounce feminism as a "counterrevolutionary diversion." Its 1969 program promised that "the Sandinista people's revolution will abolish the odious discrimination that women have been subjected to compared with men" and "will establish economic, political, and cultural equality between women and men." In actual practice, Nicaraguan women after the revolution have enjoyed greater freedom regarding the use of contraception and access to abortion, but the winning of state power has not given the Nicaraguan government full power in implementing official policies. Like all revolutionary regimes, it is hobbled by problems of material scarcity in an underdeveloped economy, by foreign-supported counterrevolution, and by various countervailing forces such as the Catholic Church, which opposes educational and family reforms, and by a sizable private sector (78 percent in industry, 76 percent in agriculture, and 60 percent in commerce) that evades legislation regulating discriminatory employment policies.[69]

So far as women are concerned, probably the chief obstacle is the deeply entrenched ethos of *machismo* rooted in centuries of Latin American culture. This was demonstrated in 1982 when the Nicaraguan village of La Estancia, with a population of seventy, was overrun by U.S.-supported contras. The surviving male villagers left their families in a refugee shelter and went off to fight in the Sandinista militia. They returned eleven months later to find their village transformed by a restored cooperative, which had put new tractors in the fields and jarringly new feminist values in the home. After languishing for three months in a disease-ridden shelter in a nearby town, the women had decided to risk returning to their farms. Sandinista soldiers tried to stop them, but the women staged a sit-down strike on the road until the military gave way. The women then rebuilt their wrecked houses, planted corn in their fields, and organized an armed patrol to protect their village. The government gave three tractors to the cooperative, and the women took courses in driving and maintaining them.

This experience in self-help and direct participation transformed the women as well as their village. "Some men don't like the idea of women driving tractors or carrying rifles," explained one of the new leaders, Feliciana Rivera. "But after struggling alone that year, we understood we had the same duties and rights as men, and we were not about to give them up." Indeed, they did not give them up. They defended tenaciously their newly won freedom, though at a high cost. They learned, as the Vietnamese women also had learned, how deeply entrenched are the "obsolete customs." No less than five marriages, including Rivera's, broke up in that village of seventy. "We had fight after fight," relates Rivera, "because I got active in the national women's association and the Sandinista Defense Committee and did volunteer health work. As far as my husband was concerned, I was doing all this to meet other men. . . . That's when I told him we'd better separate."[70]

So far as official rhetoric is concerned, the status of women throughout the world has been gaining attention and impressive support. In 1945 the United Nations charter proclaimed its commitment to "the equal rights of men and women." The U.N. General Assembly later declared the years between 1976 and 1985 to be the United Nations Decade for Women. Marking the end of that decade, a World Conference on Women was held in Nairobi in July 1985. The conference noted that "while women represent fifty percent of the world population, they perform nearly two-thirds of all working hours, receive only one-tenth of the world's income and own less than one percent of world property."

This was such a far cry from the equal rights agenda set forth in the U.N. Charter four decades earlier that the Nairobi conference concluded it was necessary for women to supplement whatever existing legal equality they might already have won with equality in political power. "Equal rights, responsibilities and opportunities in everyday aspect of life . . . can only happen if women have the means, and the power . . . to allow them to take an equal role."[71]

Power is never conferred, it is only wrested. Hence the current feminist movements to mobilize women worldwide to realize their potential power and to utilize it to win women's rights in practice as well as theory. Feminist leaders recognize the differences prevailing among their constituents scattered about the globe, but insist that they also share common problems and therefore common bonds. The crucial question, however, is whether the difference or the bonds will prevail. Will the women of the world give priority to their gender or to their national or social commitments, as did their predecessors during the French Revolution? The answer to this question remains uncertain, as evident in the current enmity between Arab and Israeli women, between traditional and liberated females in Khomeini's Iran, and between Catholic and Protestant women in Ireland.

In addition to the divisive effect of class and national differences, there is the equally deep-rooted divisiveness of clashing cultural traditions. Consider the probable reaction of an American coed to an Egyptian counterpart who was enveloped in a black chaddar, or shroud, on a hot day. When asked whether she found it uncomfortably hot, she replied, "Yes, but not as hot as in hell." Likewise, consider the Indian delegate at the 1985 Nairobi women's conference who opposed abortion and told her audience that the way to control population is "to get men to wait, to get their sexual desire under control." An American delegate shot back: "What if we want sex, honey?"[72] Despite the appeals of feminist leaders in behalf of "global sisterhood," the fact is that they confront obstacles at least as complex and formidable as those facing groups seeking to save the forests or the whales or the ozone layer.

SOCIAL RELATIONS

Over two millennia ago, Aristotle noted a connection between technology and social relations: "There is only one condition in which we can imagine not needing subordinates, and masters not needing slaves. This condition would be that each [inanimate] instrument could do its own work . . . as if a shuttle should weave of itself."[73]

If Aristotle were resurrected today, he would be astonished by the number and variety of "instruments" that "do their own work." Perhaps he would be even more astonished by the meager social impact all these instruments have had—by their failure in many respects to effect the liberation of their human creators that Aristotle originally took for granted. Of course, he would note that the slaves, so common in the Athens of his day, were gone. Yet he would be puzzled by the glaring inequities and anomalies persisting in human communities despite the obvious liberating potential of high technology.

A resurrected Aristotle would doubtless be impressed by the enormous productivity of the late-twentieth-century global economy, but sooner or later he would also be struck by the failure to reduce the work week during the past half century. Our Paleolithic ancestors (as well as today's Bushmen and aborigines) spent an average of only fifteen to twenty hours per week collecting food; at the other extreme, with the first Industrial Revolution, factory workers labored ten to sixteen hours a day, six days a week. The workday was then gradually reduced until in the United States in 1935 it was finally made forty hours a week by law. Many dreamed of further cuts to a thirty-five-hour or even a thirty-hour week with a concomitant increase in leisure activity and creativity. Instead, despite all the technological advances in recent decades, the work week in the United States has increased from an average of 40.6 hours in 1973 to 48.8 hours in 1985, and leisure time during this period has dropped from 26.2 to 17.7 hours per week.[74]

What might surprise a modern Aristotle most, however, would be the degree to which the fruits both of technological advance and economic overproduction have been unevenly distributed throughout the world, and the level of human misery that has resulted. In 1981, food consumption averaged 30 to 50 percent above normal health requirements in twenty-four countries, but in another twenty-five countries, 10 to 30 percent below requirements. The 1988 World Food Conference noted that some forty thousand children are dying each day because of illnesses related to malnutrition. In an address to that same Conference, Archbishop Renato R. Martino, the Vatican's United Nations observer, concluded that "hunger has reached such an extent and its victims are so numerous that future generations will undoubtedly regard it as the greatest catastrophe of our times, surpassing in horror and magnitude all the other tragedies that have unfortunately marked the 20th Century."[75]

In recent decades, social inequity has sharpened not only between rich and poor nations, but also between rich and poor citizens within the affluent First World, including the United States. From 1979 to 1983, for instance, the number of poor rural people in the United States jumped nearly 40 percent, from 9.9 million to 13.5 million. In New York City, the number of soup kitchens and food pantries providing emergency food increased from thirty in 1981 to five hundred in 1988.[76]

In New York City, the gap between rich and poor Americans is manifest most strikingly in what has been termed "a principality of plenty"—a narrow strip of choice New York real estate, twenty city blocks running along Park, Madison, and Fifth avenues from the Metropolitan Museum on the north to the Regency Hotel on the south. Residents in this principality have described it in terms reminiscent of Versailles and the Bourbons.

> I think there has never been such a concentration of wealth in such a small area in the history of man. . . . everybody who is really important and powerful keeps some sort of presence in New York. . . . There is probably a larger concentration of wealth than ever existed at Versailles or anywhere else. In this town, you don't have to apologize for being rich.
>
> It's far more lavish now than Paris. Paris is a very staid city now. I think London has great charm and style of life that is comparable to ours. . . . Maybe the life in New York is more opulent. We have

many more galas and grand events and black tie dinner parties.
... The food is getting better and better. We're right now in our glory.
We're at the peak of the Renaissance.

In many ways, the extremely wealthy live in the city but also
above the city. They don't cope with the same problems. They can
insulate themselves from the *Sturm und Drang* that drags down all of
us. A limousine and driver wait by the front door. Servants pick up
the cleaning, fetch the children at school, do the grocery shopping,
pay the bills. Some people have social secretaries. Their secretaries
sign their checks. The hairdresser comes to them. The masseur comes
to them.

Wealth helps to make the inconvenience and harshness of the city
evaporate. A private banker can handle financial transactions. A
family jeweler can reset important old pieces or help select new ones.
A personal shopper can circumvent lines in crowded department
stores. An art consultant or curator can supervise new additions to
collections. Accountants, lawyers, administrative assistants or financial advisers can provide distance from favor-seekers or people
pushing purchases or charities. In New York, many physicians with
wealthy practices do indeed still make house calls."[77]

In the same year that this description of New York's *beau monde*
was published, Mayor Ed Koch appointed a commission to study the
city's problems and needs. This "Commission on the Year 2000"
submitted a report that can be chillingly juxtaposed to the life-styles
of those who live in, yet "above," New York City. "Changes are
needed in a host of areas—housing, transportation, education,
health care—but reform is particularly needed in the area of poverty. Without a response to the problem of poverty, the New York
of the 21st century will be not just a city divided, not just a city
excluding those at the bottom from the fullness of opportunity, but
a city in which peace and social harmony may not be possible.
There is no more important issue for city government, no more
important test for New York."[78]

From this evidence of a widening social gap—both within and
between nations—Immanuel Wallerstein has come to conclude that
the majority of the human race is perhaps worse off today than it
was five centuries earlier in precapitalist times. The root cause for
the failure of modern technology to have a liberating effect—of
ending both want and social inequity—as Aristotle had expected is
not, as often assumed, the perversity of "human nature." Our ancestors during the Paleolithic era did not act in selfish, acquisitive, or

especially aggressive ways. The basic problem in contemporary social relations is the priority given to productivity—to the drive for profits—as against sustaining a healthy social environment. Yet the drive for profits is the essence of capitalism, the source of its inherent dynamism, which has generated the "creative destruction" that has molded the modern world.

In reaping the benefits of the "creative" component of Schumpeter's formula, we cannot escape confronting also the accompanying "destruction." This includes the social inequity and the attendant human misery. It includes also ecological depredation, so that the 1987 report of the Worldwatch Institute, *State of the World*, concludes with the significant exhortation that "the time has come to make peace with each other so that we can make peace with the earth."[79] Finally, the "destruction" entails psychological repercussions, so that 89 percent of all adult Americans in 1985 reported experiencing during the year "great stress" (symptoms including anxiety, depression, fatigue, headaches, and anger), while 59 percent said they felt such stress once or twice a week, and 30 percent reported living with high stress nearly every day.[80] Teenagers are proving as vulnerable to societal stress as adults. The number of Americans between the ages of ten and nineteen discharged from psychiatric hospitals ballooned from 126,000 in 1980 to 180,000 in 1987. An estimated 12 percent of American children under eighteen, roughly eight million in all, needed mental health services in 1989.[81]

The ambivalent legacy of capitalism explains the correspondingly ambivalent reaction it evokes. The beneficiaries of the "creativity" naturally tend to be positive, and the victims of the "destruction" to be negative. Members of the Reagan White House sported Adam Smith neckties, and proudly proclaimed their guideline to be "Enrich thyself." Ivan Boesky, of insider stock-trading fame, extended Adam Smith's "invisible hand" to its logical conclusion when he told the graduates of the University of California (Berkeley) business school: "I think greed is healthy. You can be greedy and still feel good about yourself."[82]

In practice, however, there are the greedy who do not "feel good" about themselves. They do acquire MBAs, BMWs and other symbols of success, yet find themselves strangely crippled by internal doubts and conflicts. Their predicament, according to Washington psychoanalyst Douglas LaBier, is that they internalize the often contradictory or conflicting drives that capitalism generates. He concludes on the basis of research with 230 successful career people, that they are "driven by twin motives: to be successful, competent and respected, and at the same time a desire for more fulfillment, meaning and pleasure." LaBier estimates that these "working wounded," these "victims of success," comprise over 60 percent of the total career force. "There's a pervasive depression and sense of helplessness in the career culture," concludes LaBier, "and its not because people have messed up brain chemicals or miserable childhoods. There's something in our culture that generates depression as a by-product."[83]

What psychoanalysts have found in working with individuals, sociologists have found in studying communities. In *Habits of the Heart,* a team of social scientists recently examined the state of American society and its long-term viability. Their main finding is that modern technology has had "devastatingly destructive consequences" not only for our "natural ecology" but for our "social ecology" and "moral ecology." "What has failed at every level— from the society of nations to the national society to the local community to the family—is integration. . . . we have put our own good, as individuals, as groups, as a nation, ahead of the common good.

> . . . we have never before faced a situation that called our deepest assumptions so radically into question. Our problems today are not just political. They are moral and have to do with the meaning of life. . . . Perhaps the truth lies in what most of the world outside the modern West has always believed, namely that there are practices of life, good in themselves, that are inherently fulfilling. Perhaps work that is intrinsically rewarding is better for human beings than work that is only extrinsically rewarded. Perhaps enduring commitment to those we love and civic friendship toward our fellow citizens are preferable to restless competition and anxious self-defense. Perhaps common worship, in which we express our gratitude and wonder in the face of the mystery of being itself, is the most important thing of all. If so, we will have to change our lives and begin to remember what we have been happier to forget.
>
> We have imagined ourselves a special creation, set apart from other humans. In the late twentieth century, we see that our poverty is as absolute as that of the poorest of nations. We have attempted to deny the human condition in our quest for power after power. It would be well for us to rejoin the human race.[84]

The invitation to "rejoin the human race," however, may only be an invitation to leave the frying pan for the fire. The fact is that "creative destruction" has disrupted not only American society but all human societies. The resulting stress and discomfort have triggered even sharper debate about private gain versus the common good in the socialist Second World and the underdeveloped Third World than in the capitalist First World.

In China, during Mao's Cultural Revolution, one-fifth of the human race was relentlessly brainwashed with the ubiquitous slogan "Serve the People, Fight Self." Mao's basic tenet has been that every revolution bears within itself the seeds of counterrevolution. Postrevolutionary societies require the services of political experts (party and state cadres) and economic experts (managers, technicians, and scientists). If these experts are favored with special financial and political rewards, then the end result is Soviet-type meritocratic elitism that Mao regarded as the antithesis of socialism. Hence, Mao insisted that every expert should also be "red"— that is, a "comrade" who did manual as well as intellectual work, and who focused on serving society rather than self.

An example of Maoist indoctrination was the propaganda campaign based on the life of a young soldier, Lei Feng. During his short life before he was killed in an accident, Lei Feng was portrayed as a model comrade-citizen. He kept his army truck in perfect condition, contributed his meager pay to the aid of flood victims, worked in his spare time on public projects, and always avoided acclaim or reward. The press emblazoned his good deeds and a film was made of his life. A foreign correspondent who saw the film noted that the audience left the theater wet-eyed. When he ventured to suggest that the sermonizing was overdone, he was met with utter incomprehension. After all, Lei Feng really did exist, and he really did perform all those good deeds, and there are many other Lei Fengs in China, "and we can congratulate ourselves on this; it is only with such men that we can advance toward the collectivist society."[85]

Mao's successor Deng Xiaoping reversed China's priorities, placing expertness before redness. Government slogans no longer called on the people to "Fight Self"; rather, they assured citizens that "To Get Rich Is Glorious." Consumer expectations rose correspondingly—from the "big four" (bicycle, sewing machine, radio, and watch) to the "big six" (color TV, washing machine, radio-cassette player, refrigerator, electric fan, and motorcycle) and most recently to the "eight big things" comprising the "big six" plus modern furniture and a camera. Students formerly had been warned against the "private ownership of knowledge"; that is, against using their education to advance private rather than community interests. Under Deng ambitious students are learning foreign languages and contriving by any means to go abroad to study. Eighty-five percent of those who go to the United States remain there, where the approved wealth and glory are more easily attainable.[86]

Under these new circumstances, Lei Feng has become a quaint figure from the past. The mention of his name now provokes scornful laughter. A young man in a marketplace dismissed Lei Feng as irrelevant when asked about the former Maoist hero: "In the past few years people put money first. No one wants to live a poor life. No one cares about these political slogans. What the people care about is how to improve their lives. We're not fools."[87]

Yet by no means do all Chinese embrace this emphasis on individual advancement. Some affluent young entrepreneurs are finding it difficult to win brides because their newly acquired wealth is widely considered to be improperly obtained and compromising. At a more abstract level, *China Youth News*, the country's preeminent publication for young people, published on the anniversary of Lei's death a poignant essay seeking with evident confusion and unease to resolve the conflict between serving self and serving society. "Self-realization is not and should not be the ultimate purpose of life. Even if one's ability could develop perfectly, it would finally die with the end of one's life. In fact, there is no such thing as an everlasting value that has absolutely nothing to do with society and that is utterly of a personal nature. Only when a value belongs to mankind can it be everlasting." From this preliminary observation, *China Youth News* deduced that Lei was fully justified in devoting his life to unreserved service to the people. "This is the only correct way to realize personal value." The majority of Chinese youth, stated the journal, "hope that the general mood of society will turn for the better, that unhealthy practices in the Party and bureaucratism in state organs will soon be eliminated, that production departments will produce good quality products, that service and commercial units will improve their services. . . . [But] how can all this come about," asks the journal, "if publicity is given to 'self-realization' rather than to 'serving the people'?"[88]

The crisis of confidence reflected in this ongoing public debate in China is by no means peculiar to that country. Soul-searching is at least as widespread and as intense among Soviet citizens now being subjected to Gorbachev's unceasing exhortations for new thoughts and new actions. Likewise, in Yugoslavia, a "senior" official has expressed skepticism and defeatism to the point of total abdication. "Our hope is with the young generation. They are skeptical of words, they understand the world. They are aware of the deficiencies of all systems. Today's divisions are not classical, not socialism or capitalism, West or East. They are wiser than we; we are slaves of the past."[89]

WAR

The nature of war was transformed as fundamentally by capitalism in modern times as it had been earlier by tributary civilization. In feudal Europe, warfare had been dominated by mounted knights. In the eleventh century, for instance, a few hundred Norman knights were able to conquer and rule southern Italy and Sicily, and later wrest the Holy Land from the Moslems. In the twelfth and thirteenth centuries, the military scene was transformed by Italian city-states using their commerce-based wealth to recruit soldiers to man their city walls, and to arm these recruits with pikes and crossbows to withstand charging knights in open fields. As the ranks of pikemen and crossbowmen came to be supplemented by and coordinated with cavalry for flank protection, counterattack, and pursuit, the art of warfare became more complex.

By the fourteenth century, such coordinated forces had evolved into independent bands of professional troops commanded by captains who negotiated contracts with Italian city officials for specified military services for specified periods. These contractors, or *condottieri,* headed bands as large as ten thousand armed men, and the contracts they signed and fulfilled represented the commercialization of armed violence. Officials of wealthy Italian towns taxed their citizens to pay for military protection because it was cost-efficient. "Ravaging soldiers," states William McNeill, "no longer had to sustain themselves by forcibly recirculating the movable wealth of a country. Regular, predictable taxes did the trick instead, transferring money from civilians to officials who used it to support an efficient military force as well as themselves."[90]

The military prototypes evolved by the Italian city-states were soon adopted and improved by the large and wealthier new northern European states: France, England, and the United Provinces. With their superior resources, they assembled large standing armies that were well armed, well trained, well disciplined, and well organized under a clear and undisputed chain of command stretching from crown to officers to rank-and-file. The spectacular victories of these European military establishments in overseas lands is commonly attributed to the technological superiority of European firearms; but at least as decisive was the superior steadfastness and maneuverability of the thoroughly drilled and disciplined European troops. First, they overran the vast underpopulated regions of the Americas and Siberia. The booty from the new colonial possessions facilitated the financing of still larger military and naval forces, which enabled the Europeans to extend their empires into Asia and Africa. Military, political, and economic expansionism proceeded in a self-reinforcing cycle. Transcontinental railways, intercontinental canals, and steam-powered warships allowed Europeans to transport and concentrate their forces at will throughout the world, and to pick off any parcel of real estate that caught their fancy. During Queen Victoria's reign, usually thought of as a time of peace and tranquility, the British waged no fewer than seventy-two separate campaigns globally—more than one per year.

In the late nineteenth century, European military technology effected a great leap forward with an unprecedented integration of science and the military, resulting in what William McNeill has labeled "invention on demand." This development occurred first in Britain, and was initially focused mainly on naval armaments. Heretofore, individual inventors had led the way in originating new weapons and then had sought buyers in the armed services. Now the process was reversed, with military technicians setting the goals and specifications of projected new weapons, and private firms competing for contracts to produce them. Previously, government arsenals had been the principal arms manufacturers, but now they were steadily supplanted by private companies like Armstrong in

Britain, Krupp in Germany, and Schneider-Creusot in France. These companies invested in steel mills, shipyards, and heavy machinery far superior to the facilities available in government arsenals. Their soaring costs were covered by weapons sales to foreign governments, so that private companies enjoyed advantageous economies of scale unavailable to arsenals producing only for their respective governments.

The private companies were also aided by the extension of the franchise in western European countries at the turn of the century. Newly enfranchised voters generally supported larger government expenditures for arms expenditures that helped to generate jobs and relieve unemployment, particularly because they were assured that the rich would be forced to foot the arms bills through heavier taxes. Politicians now were willing to vote for increased military appropriations because such votes were becoming assets rather than handicaps at the polls.

The convergence of interests among public officials, military officers, and armament company executives, as they moved back and forth across new financial bridges connecting the business and public realms, led by the 1890s to what is now known as the "revolving-door phenomenon." By the turn of the century, what U.S. President Dwight D. Eisenhower later called the "military-industrial complex" had emerged in Europe.

A powerful feedback loop established itself, for technological transformations could not have proceeded nearly so rapidly if economic interest groups favoring enlarged public expenditure had not come into existence to facilitate the passage of bigger and bigger naval bills. Each naval building program, in turn, opened the path for further technological change, making older ships obsolete and requiring still larger appropriations for the next round of building. . . . for thirty years, 1884–1914, it [command technology] grew like a cancer within the tissues of the world's market economy.[91]

The "cancer" metastasized from national to global proportions because sales to one foreign country aroused fears and stimulated sales to neighboring countries, triggering a worldwide arms race, manipulated by arms salesmen who were castigated by contemporary critics as "merchants of death."

If these pre–World War I developments seem disturbingly familiar today, even more disturbing was the direction in which they led. The high commands of all the major powers coordinated their new military technology with their dense railway networks to prepare detailed mobilization plans by which huge quantities of military supplies and millions of conscripted soldiers were to be moved to war fronts against any conceivable combination of enemies. Once a mobilization order was given, men and supplies were programmed to move like clockwork and on a massive scale. No countermanding order was considered tolerable, whatever the political situation, because the flow of the military machinery would be jammed and the country left defenseless in the resulting bedlam. In a crisis situation, the ponderous preset plans made it almost impossible for a major mobilization order to be recalled or throttled down by any authority, whether emperor or prime minister. This "exigency of the military timetable," as historians have dubbed it, proved to be a major factor in the breakdown of the efforts to reach a peaceful settlement following the Sarajevo assassinations on June 28, 1914. At least the diplomats of that era had a grace period of several weeks before guns actually began firing on August 4. Today the military timetable has shrunk almost to the vanishing point with modern weapons-delivery technology able to span continents in minutes, and relying on computers, as it does, leaving virtually no time for human participation in the decision to unleash a nuclear Armageddon.

During World War I, the number of weapons appearing on "demand" increased exponentially. Experience in battle prompted demands for rapid improvements in existing weapons such as U-boats and airplanes, and led to the invention of new weapons such as tanks and poison gas. So rapid was the pace of technological innovation that before peace broke out in November 1918, the British High Command approved "Plan 1919" which was to utilize a new, advanced generation of speedy tanks that would break through enemy lines and continue operations in the enemy's rear areas. The war ended before the new weapon and strategy could be tested. The Germans refined both during the interwar years, and unleashed them with devastating effect in 1939 in the form of its now famous blitzkrieg campaigns.

What happened during World War I was repeated during World

War II, at an accelerated pace and at a higher level that reflected the unceasing technological advances. Again, weapons appeared on demand, and on a scale that laid waste whole continents and left 50 million dead, more than twice as many as in World War I. American financial and human resources were mobilized for the Manhattan Project, which engaged four thousand scientists, including fifteen Nobel laureates, and which generated the first two atomic bombs, promptly dropped on Hiroshima and Nagasaki and killing an estimated 210,000 civilians.

The German counterpart to the nuclear breakthrough was the V-2 rocket that ravaged London and later spawned the numerous ballistic missile systems now proliferating throughout the globe. The Germans went further in industrializing and bureaucratizing death as a factory process with their extermination camps operated as assembly-line death factories. These were designed to achieve Hitler's "final solution" of the Jewish problem, as well as to clear the eastern European "lebensraum" of the inferior subhumans *(Untermenschen)* who were to be replaced by superior Nordics. Five large extermination camps were constructed, of which Auschwitz achieved a grim record for efficiency in production-line murder— twelve thousand per day. Thus perished an estimated 6 million Jews, along with 5 million Protestants, 3 million Catholics, and half a million gypsies.

Such wholesale slaughter was not unique in human history. Nomadic invaders from the central Eurasian steppes left countless victims in their wake, as did also the fanatics waging their interminable religious wars, and the western Europeans expanding overseas in search of gold and heathens, and in the process wiping out defenseless native populations. But as Arnold Toynbee has noted, twentieth-century genocide is unique in "that it is committed in cold blood by the deliberate fiat of holders of despotic political power, and that the perpetrators of genocide employ all the resources of present-day technology and organization to make their planned massacres systematic and complete."[92]

The Nazis did indeed make full use of their technological resources. They used their prisoners as a source of labor while they were alive and as a source of "raw materials" after death. They ordered the ashes in the ovens carted off to be used as fertilizer, the hair from the corpses used for mattresses, the bones crushed for phosphates, the fat used to make soap, and the gold and silver fillings from teeth deposited in the vaults of the Reichsbank.

Albert Einstein quickly sensed the implication of these evils nourished by the Second World War. In May 1946, he issued his famous warning that "the unleashed power of the atom has changed everything except our ways of thinking, and thus we drift to unparalleled catastrophe."[93]

Einstein's warning has been ignored because the full flowering of the "military-industrial complex" following World War II has institutionalized "invention on demand." In late nineteenth-century Britain, naval contracts were supported by newly enfranchised voters fearful of unemployment. In the United States, military production expanded from about 1,600 federally-owned plants in the 1940s to over 30,000 private companies operating in every region of the country in the 1980s. These companies negotiate over 15 million contracts each year with various government military agencies. Preparing for war has become both profitable business and successful politics, as evident in the typical American electoral contest in which all candidates pledge to support new weapons systems. A super weapons complex like the Strategic Defense Initiative (SDI, or "Star Wars") highlights the scope of the prevailing military-industrial complex. Congress began authorizing SDI appropriations in 1983, and by 1987 contracts had been signed with eighty universities (led by MIT with $350 million) and with 460 corporations (led by Lockheed with $1 billion). SDI's impact extends well beyond the United States, as this 1985 comment by Canadian Prime Minister Brian Mulroney indicates: "I suppose, if somebody came forward and said, would we be interested in bidding for a part of a [SDI] contract which would create, say, 10,000 jobs in Winnipeg-Fort Garry, I think we would have to take a look at it."[94]

Under these circumstances, the torrent of "demand" inventions continues unabated. The atomic bombs that appalled Einstein in the 1940s now serve merely as triggers for modern hydrogen bombs. Just as in 1906 Britain's *Dreadnought* triggered a naval armaments race between Britain and Germany, so the 1945 American A-bomb was followed by a Soviet one in 1949; the 1952 American H-bomb by a Soviet one in 1953; and the 1968 Soviet antiballistic missile by an American one in 1972. The result of this arms race is the accumulation of a global arsenal of fifty thousand nuclear weapons. Such developments have hardly abated even in the face of a warning issued in November 1983 by a consortium of scientists from several countries that if only a small fraction of existing nuclear weapons were detonated, it would precipitate a "nuclear winter." Firestorms and massive amounts of smoke, oily soot, and dust would blot out the sun and plunge the earth into a freezing darkness for from three months to a year or more. "Global environmental changes sufficient to cause the extinction of a major fraction of the plant and animal species on the earth are likely. In that event, the possibility of the extinction of *Homo sapiens* cannot be excluded."[95]

The unprecedented evil perpetrated by Hitler against the Jews prompted the coining of the term "genocide," from the Greek *génos* ("race" or "nation") and the Latin *cide* ("killing"). Today's prospect of an even greater evil has given birth to another new term, "omnicide," connoting the murder not of a nation but of the human species.

How paradoxical is the current foreboding of the possibility of the extinction of the human species at the very time of its greatest triumphs, when virtually anything imaginable by the human mind seems to be attainable by human technology. The root cause for this paradox is manifest in the millions of years of human development from *Australopithecus* to *Homo sapiens*, as reviewed in this book. The course of this development discloses that full-scale warfare, as distinct from personal vendettas, had its roots in society rather than in genes. Humans did not engage in warfare until they began accumulating possessions worth fighting over, and that did not happen until very recently in the overall course of human history. It did not, in fact, happen until the Agricultural Revolution and the Industrial Revolution increased human productivity dramatically, creating material societies, or civilizations, fundamentally different from the subsistence societies of all preceding food-gathering peoples.

These civilizations were based on cultivated fields, overflowing granaries, and urban centers filled with manifold treasures, all valuable prizes for predators within and without. Only then did war became profitable and endemic, waged by nomads irrupting out of their deserts and steppes, by Roman senators seeking new provinces to loot, by conquistadors overrunning whole continents with their muskets and crosses, and by their successors who achieved dominion over worldwide empires with their gunboats and machine guns, and most recently with their helicopters and computers.

The great paradox of our age is that just as the Agricultural and Industrial revolutions made war profitable and "rational," so the ongoing High-Tech Revolution has made it unprofitable and suicidal. The technological precociousness that enabled humans to create their own environments rather than await genetic adaptations as have all other species, has now led us into a novel environment to which humans must adapt or perish. Einstein was merely recognizing this cardinal fact when he warned that humanity now faces the choice of new "ways of thinking" or "unparalleled catastrophe."

Innumerable species have disappeared in the past because of a sudden environmental change like the Ice Ages, to which their genetic mode of adaptation was too slow to adjust. Humans are uniquely different because they have been able thus far to use their brains to create their own environments (by technologies such as the use of fire, the making of clothes, and the building of shelters) to suit their own needs. Now they are confronted with the latest of these human-made environments, one evolving rapidly and requiring quick adaptation. Failure to meet this requirement seems likely to end with a familiar pattern of ossification as the prelude to disappearance. Up to now, the human brain's ability to mold suitable environments has made the human epic the great success story of planet Earth. The supreme question now facing humans is whether they can once more use their brains adequately, this time to adapt expeditiously to a new world of their own creation.

In the final analysis, the problem of adaptation is a problem of values. This has been true for millennia, as each technological revolution generated social conflicts that raised questions involving values. The Agricultural Revolution and the ensuing class-based civilizations with their landlords and landless prompted numerous religious leaders throughout the world to preach in support of social welfare and against individual covetousness. As the Acts of the Apostles testify: "The group of believers was one in mind and heart. . . . Those who owned fields or houses would sell them, bring the money received from the sale, and turn it over to the apostles; and the money was distributed to each one according to his need."[96] Likewise, the Scientific Revolution and the resulting quantum jump in human productivity and power prompted the pioneer scientist Francis Bacon to warn that the new knowledge should be used not for individual "profit or fame," but for "charity . . . and benefit of life."

Bacon's warning is as valid today as it was when uttered in the early seventeenth century. The basic issue of values remains unchanged through the ages except for one qualitative difference—a difference in urgency. Humanity has been able for millennia to ignore its prophets and yet manage somehow to muddle through, bloodied but surviving. The potency of modern high technology has abolished this comfortable time cushion. Psychiatrist Robert Jay Lifton has observed that when the question of American-Russian relations is raised, "the answer is not ideological, it is really pragmatic—the answer is, 'If they die, we die. If they survive, we survive.' Shared fate is the beginning of something . . . I call the species self—a sense of self in the true psychological sense of being bound up with every other single self, individually and collectively, on the globe. The beginnings of that growing awareness have been forced upon us by our technology of destruction. . . . I think that is an important psychological process to cultivate. A sense of shared fate and of the species self, psychologically and politically, is an idea whose time has come."[97]

Indeed its time has come. What is on trial today is not any one race or nation or "ism." It is humanity itself—the ironically self-styled *Homo sapiens.*

CHAPTER 4

HUMAN
PROSPECTS

We can only sound the alarm, again and again; we must never
relax our efforts to rouse in the peoples of the world, and especially
in their governments, an awareness of the unprecedented disaster
which they are absolutely certain to bring on themselves unless
there is a fundamental change in their attitude toward one another
as well as in their concept of the future. . . . The unleashed power
of the atom has changed everything except our ways of thinking,
and thus we drift to unparalleled catastrophe.

ALBERT EINSTEIN (1946)

A NEW AXIAL AGE

If humanity is on trial today, then this represents a perplexing tangling of the lifelines from our past. Ours is a period of unprecedented human achievement, whose greatest success has been precisely in an area—productivity—where inadequate performance has been the root cause of the demise of two forms of social organization, kinship and tributary societies. Those earlier systems had emerged in response to human needs, and both had prevailed for extraordinarily long periods because they had fulfilled those needs satisfactorily. Both systems, however, proved incapable of meeting the material demands of growing human populations, and each gave way to a new social organization better suited to coping with new needs. Hence, the replacement of kinship society by tributary, and then of tributary by capitalist.

Capitalism's ruling principle, profit or perish, gave it an intrinsic dynamism that persists to the present. From its origins in northwestern Europe in early modern times, it has evolved through three stages, from Commercial to Industrial to the current High-Tech Capitalism, each being more productive and expansionist than the last. Thus, capitalism has inexorably broadened the base of its operations from local to national to global. The world we know is a world made by capitalism, a combination of unrelenting and ever-increasing productivity, together with its inescapable corollary, unrelenting and ever-increasing consumerism. This is a combination that is now so globally pervasive and so taken for granted that it seems almost inconceivable not to consider it an unavoidable manifestation of human nature. Yet it is, in fact, a phenomenon of remarkably recent human vintage—an expression not of the innate impulses of *Homo sapiens* but of modern capitalist dynamism.

During the Paleolithic millennia, the accumulation of private possessions was not only unrealistic because of band mobility, but also eschewed because of band mores which stigmatized acquisitiveness as an intolerable deviation from propriety. By contrast, conspicuous consumption was all too prevalent under the following tributary civilizations, as evident in the vast quantities of luxury goods buried in the Pyramids and other graves of the powerful and affluent worldwide. But this was only a limited elite consumerism, in contrast to the mass consumerism made possible and imperative by the productivity of modern capitalism.

In preindustrial Europe, 60 to 80 percent of individual income was spent on food, leaving little for anything else. A new garment was such a luxury that even the clothes of plague victims were eagerly sought after by relatives. The situation was similar in the rest of the world, which explains why the Japanese kimono remained unchanged for centuries, and likewise the pajama in the Middle East, the dhoti in India, and the poncho in pre-Columbian America.

Nowhere could masses of people indulge in consumerism until the Industrial Revolution generated mass productivity, which in turn necessitated mass markets and an acquisitive nature in the population at large. This was recognized in the eighteenth century by James Watts's financial partner, Matthew Boulton, when he observed: "We think it of far more consequence to supply the People than the Nobility only; and though you speak contemptuously of Hawkers, Pedlars, and those who supply Petty Shops, yet we must own that we think they will do more towards supporting a great Manufactory, than all the Lords in the Nation."[1]

Businessmen soon recognized the need not only to search for customers among the masses, but also to persuade those masses that it was in the nature of things to need to purchase certain commodities hitherto ignored or previously unknown. As early as the eighteenth century, a full range of what are now considered modern selling techniques, including market research, credit, discount schemes, handbills, catalogues, newspaper and magazine advertising, and money-back-if-not-satisfied sales offers had been developed. A pioneer in this incipient mass merchandising was the potter

Josiah Wedgwood, who stated candidly that "Fashion is superior to merit." Accordingly, he conducted sales campaigns that made his pottery the best known and most desired in the world, though often it was neither the best nor the cheapest.

Mass-merchandising techniques did succeed in creating a mass market. Customers were persuaded to regard what had once been almost unimaginable "luxuries" as "decencies," and later as "necessities." The process accelerated when the second Industrial Revolution generated floods of new commodities that had to be marketed in one way or another. Between January and April 1987, to take an almost random example, no less than 3,152 new food, household, or beauty items made their debut on the shelves of American supermarkets—one new product every forty-one minutes. To house this flood of new items, American supermarkets are becoming more super each year. In 1989 construction of the world's largest shopping center began in Bloomington, Minnesota, a suburb of Minneapolis. This "megamall" will cover seventy-eight acres and include thousands of shops as well as a hundred nightclubs and restaurants, eighteen theaters, a miniature golf course, and a roller coaster that will roar through the middle of the mall.[2]

The power and dynamic appeal of capitalism, its ability to practically unravel other types of societies, is evident in China where only a few years ago Mao had sought to create an alternative social order based on the common good rather than private profit. His prospects seemed promising, given the sheer size of the Chinese nation, its historic autonomy on the eastern end of Eurasia, and the impressive power and initial popularity of the Communist revolution. Yet today, only a relatively few years after Mao's death, the most basic of his principles and practices are being casually abandoned or reversed.

This is well illustrated at the moment of writing by what is happening in China's Guangdong province opposite the British colony of Hong Kong. In a major policy shift, the Chinese government decided in 1988 to open its entire coastal area to foreign investment, which hitherto had been restricted to fourteen coastal cities and four Special Economic Zones. The new Coastal Development Strategy, as it is called, opens to foreign investment coastal territories with a population of 200 million. Foreign corporations are free now to bring into these territories their own management methods, labor-intensive technology, and raw materials, limiting China's contribution to an abundant supply of cheap labor.

In this context, Wisegroup Investment Ltd., based in Hong Kong, has established a knitted-garment factory in the town of Shenzhen just across the colony's border. Its sixty-two employees work ten hours a day, seven days a week, for thirty cents an hour. "Everyone wants to come here," explains Li Laikam, a twenty-year-old worker who moved here from her native village 260 miles to the southwest. "It's better here because you don't have to endure the wind and the rain in the fields." How many times similar testimony must have been heard in eighteenth-century England when farm families were leaving their villages for the new factories sprouting in Manchester, Leeds, and Birmingham.

Such capitalist economic intrusion into China is paralleled by a disruptive cultural intrusion, symbolized by the readiness of some Chinese women who have adequate funds to spend them on cosmetic surgery so that they might resemble the Western or Westernized models so admired in new magazines and on new television shows and ads. In a country where a typical monthly salary is under thirty dollars, they are paying seven to eighty-five dollars for nose reconstruction, six to thirty dollars for eyelid operations, and up to nine hundred dollars for breast augmentation, and making a success of companies producing over-the-counter creams that promise to lighten skin color.[3] Not too long ago the ancestors of these young women were referring contemptuously to the Western newcomers as "long-nosed barbarians"! Those ancestors doubtless would be equally shaken by the five-hundred-seat "Kentucky Fried Chicken" that has opened on a corner of Beijing's historic Tiananmen Square, with a view of the Gate of Heavenly Peace and Mao Tse-tung's mausoleum. The new enterprise retains its familiar slogan in Chinese—*Haodao shun shouzhi*—"So good you suck your fingers."[4]

In addition to Kentucky Fried Chicken, Chinese citizens now can enjoy the game Monopoly, two hundred thousand sets of which have been manufactured and sold by Shanghai's Lishen Toy Factory, under the name Strong Hand. Parker Brothers, the American owners of Monopoly, were scooped by the Lishen Factory, which was not authorized to produce and market Strong Hand. So Parker Brothers, determined not to be scooped again by a socialist company, has announced that it will introduce a Russian-language version of Monopoly in the Soviet Union.[5]

Capitalism's global cultural impact is not confined to underdeveloped Third World countries. Only a few years ago, French leader Charles de Gaulle withdrew France from NATO, and French cultural ministers were denouncing American "cultural imperialism." Yet France now is welcoming back Mickey Mouse and Donald Duck. The Walt Disney Company has negotiated a contract with the French government to construct a "Euro Disneyland" near Paris. The exceptionally favorable terms granted to the American corporation have been defended by French officials on the ground that Euro Disneyland will make the Paris region "the most formidable pole of attraction in Europe." At the same time, at the other end of Europe, the Russians are adopting the American lingo of jazz, rock, and consumerism through their young West-oriented hipsters and black marketeers. Even Gorbachev's *perestroika* is stimulating the borrowing of American business terminology. His new world of restructured communism is populated by *biznessmeni* in search of *sponsori* to help *finansirovat* their *kooperativi,* and to organize a *dzhoint venchur* with Western *partnyori* who will contribute *tekhnologiya,* especially *kompyuteri.*[6]

Such capitalist inroads throughout the globe suggest an assured future for global capitalism. If kinship and tributary societies prevailed for millennia because they satisfied contemporary human needs, then the same criterion should indicate that modern capitalism, a mere half millennium old, is still only in its infancy. This seems especially true since the world of the late twentieth century has solved with a vengeance the basic problem of inadequate productivity that undermined preceding human societies. The annual economic summit conferences of the leaders of the major industrial powers have been concerned above all else with the dilemma of global glut, not global scarcity.

Yet at this point we come upon a great paradox of our age—capitalism is now being questioned and challenged throughout the globe precisely because of its rampaging success. Its new heights of creative energy have been accompanied by new heights of destructive energy, an energy manifest throughout the globe. It is manifest in the want amid plenty among nations, in the want amid plenty within nations, in the mindless wastage of natural resources and the equally mindless damaging of the physical environment, in the widespread absence of a sense of well-being in the present, and in the widespread apprehension not just about the future but about whether there will be a future.

This all-pervasive paradox stems directly from the way capitalism's creativity and destructiveness, noted by Schumpeter decades ago, seem welded together. The question today, when both impulses have reached unprecedented proportions, is whether new material and human conditions exist that would allow for the curbing of the destructiveness without the blunting of the creativity—whether some way can be devised by which the wondrous technology can be harnessed to serve long-term human needs rather than the reverse. Put in this way, the problem confronts all peoples—not only those of the capitalist First World, but also those of the socialist Second World who are now debating whether they must compromise with capitalist principles to cope with their chronic underproductivity, and also those of the Third World who are searching desperately for some way out of their historic subservience to the global capitalist economy.

If the human dilemma today is basically one regarding values in the most basic sense, then it is not unique to our time or society. All the great religions and their prophets are replete with exhortations regarding values. Indeed, there is a significant parallel between the circumstances in which the great religions arose and the circumstances prevailing today. Some two to three millennia ago when those religions were taking shape, a contemporary high technology was triggering a leap forward in productivity with social repercussions somewhat comparable to what is going on now. The new materials, communication facilities, and power sources of our age had their counterparts in the first millennium B.C. in the invention of the alphabet, of coinage, and of iron metallurgy. The latter invention enabled farmers to use sharp and durable iron axes and iron-shod plows to extend agriculture beyond the restricted river valleys and adjacent uplands where it had hitherto prospered. They cut down heavy forests previously invulnerable to stone-edge axes and wooden plows. They extended agriculture southward from the Yellow River valley, eastward from the Indus River valley, and from the Middle East westward to central and northern Europe, and eastward across the Iranian plateau.

The dramatic extension of agricultural frontiers resulted in a corresponding increase in agricultural output. The surplus now available stimulated trade and supported craftsmen who provided the goods and services needed in the growing economy. Those were the centuries when caravan routes were opened up across the interior of Eurasia, and sea routes crisscrossed the circumference of the Eurasian landmass—from the North Sea to the Levant, from the Red Sea to India, and from India to Southeast Asia and China. The upsurge in commerce was facilitated by the invention of coinage, which replaced the awkward barter system hitherto prevailing. The net effect was a stimulus for local and interregional commerce, a corresponding stimulus for handicrafts and agriculture, and an overall increase in economic specialization and productivity.

These economic developments led, in turn, to social and political changes. The military aristocracy that had risen to prominence with the nomad invasions of the second millennium B.C. was now displaced in many regions by a rising new class of merchants, craftsmen, and mariners. The old tribal society was transformed by monetization. Personal services and allegiances were superseded by marketplace considerations. Tribal chiefs and their advisory councils and assemblies were displaced by kingdoms and then by empires.

These developments involved profound changes in social relationships, in political activities, in ways of living and of earning a livelihood. Such all-inclusive disruption was unsettling and uncomfortable. It threw into question the nature of human existence. It led to soul-searching—to the posing of new questions and the seeking of new answers. Thinkers were moved to reconsider their respective traditions and either to abandon them or to adapt them to meet the needs of a changing world. Philosophers such as Plato and Confucius traveled from court to court as advisers and teachers, desirous of training the statesmen of the future.

Throughout Eurasia—from China to India to the Middle East to the Mediterranean basin—the intellectual ferment brought to the fore questions about the moral basis of ideal government, the functioning of the social order, and the origins and purpose of the universe and of life. Answers varied from region to region in accord with local historical traditions and objective conditions, but the range of answers constituted the great philosophical, religious, and social systems of the first millennium B.C. It was not happenstance that the spokesmen for all those systems—Confucius in China, Buddha in India, Zoroaster in Persia, the prophets in Palestine, and the rationalist philosophers in Greece—appeared during that millennium.

At a time of social unrest and moral confusion, the new religions offered solace, security, guidance, and sometimes new and alternative ways of organizing human life. They offered salvation—an afterlife of eternal bliss. They offered brotherhood that was open to all who sought admittance, women as well as men, rich and poor alike, slave or free. They stressed a high code of ethics, the observance of which was essential for salvation, and which influenced decisively the daily lives of the faithful. So profound and lasting was the impact of the new philosophies and religions that these centuries have been labeled the "axial period" of human history.[7]

Given the parallels with what is going on today, it would not be unreasonable to speculate that the late twentieth century may some day be viewed as a new axial age. Our technological revolution is perhaps deeper, as are also its social repercussions. Its scale is global, rather than being limited to certain regions of Eurasia, and the masses of the people, thanks to modern communications, are activated and assertive to an unprecedented degree. So again, it is unlikely to be happenstance that new religions, new philosophies, new social movements, and new leaders are beginning to crop up on every continent. As in the original axial age, basics are being challenged—governments, isms, traditions, and leaders. In the course of today's axial age, one and all are now on trial.

TOWARD ALTERNATIVE CAPITALISMS

Even though the world of the late twentieth century is dominated economically and culturally by capitalism, the capitalist First World finds itself under critical attack, both at home and abroad, as does the rival socialist Second World. The critique in the First World itself is muted, however, because global capitalism is still basking in the warm afterglow of the golden boom years following World War II. While the 1980s were not quite as prosperous and problem-free as the preceding decades, the spirit remained, by and large, generally upbeat, especially in elite circles. Whereas self-criticism and restructuring plans in the Soviet Union originate with Gorbachev and his colleagues in the Kremlin, the prevailing mood in the White House and in corporate suites has been one of self-satisfaction rather than self-criticism—a mood in which leaders are averse to tinkering with anything that is not obviously broken, and when the breakdown is obvious, the reaction tends to be to avert the eyes as long as possible. An example of this confident spirit is afforded by the July 1987 seventieth-anniversary issue of the business magazine *Forbes*, the self-styled "capitalist tool." Contributors to this issue note that when Joseph Schumpeter died in 1950, his teachings were eclipsed by those of Keynes, who preached government-directed fine-tuning of the economy. Today the situation is reversed, with entrepreneurs, "the agents of painful change," everywhere in the saddle, while "bureaucrats and stay-put managers are suspect." The only remaining world problem of significance, from the *Forbes* purview, is how to curb well-meaning democratic governments that try to "protect the victims of change," and in the process stifle capitalist creativity and progress. "Do we sincerely want creative destruction? Or will the government move in to stop it?"[8]

The answer of *Forbes*'s contributors is exuberantly obvious. They quote economist F. M. Sherer of Swarthmore College, who states that we are on the threshold of a new outburst of Schumpeterian inventiveness after a lull in the late 1960s and 1970s: "With the ability to synthesize DNA, the many manifestations and continuation of the microelectronics revolution and the third main component, superconductivity, you've got the making of a fifth Schumpeterian Kondratieff cycle. We are at the turning point of an upturn."[9] Also quoted is Robert White, president of the National Academy of Engineering, who sees information technology merging with biotechnology and other fields to create new resources for products and processes: "These areas will be as sweeping in their implications as anything in the past two centuries. It's only a question of time."[10]

A peak of confidence is reached by former Citicorp Chairman George Moore, who predicts that in 25 years the world will be dominated by a handful of large financial institutions and they could have their headquarters on space platforms to avoid government regulations."[11]

An escalating global domination by multinational corporations, however, is a prospect that is opposed by many of those targeted for that domination, as will be noted in the forthcoming section on the Third World. Rejection and opposition may be mounting even within capitalist homelands. This challenge to a multinational-dictated future is multifaceted, being based on considerations partly ethical, partly ecological, and partly social in nature. The ethical aspects of the challenge have been laid out most clearly in the Pastoral Letter adopted in November 1986 by the U.S. National Conference of Catholic Bishops under the title *Economic Justice for All: Pastoral Letter on Catholic Social Teaching and the U.S. Economy.* The Pastoral Letter rejects the assumption that an "invisible hand" operating through the marketplace can best settle the major domestic and international economic problems. Instead, the bishops presented an alternative humanistic approach:

The market is limited by fundamental human rights. Some things are never to be bought or sold.

The Church's teaching opposes collectivist and statist economic approaches. But it also rejects the notion that a free market automatically produces justice.

Competition alone will not do the job. It has too many negative consequences for family life, the economically vulnerable and the environment.

The bishops then applied these general principles straightforwardly to specific issues. Regarding work, they defined it not merely as something to be endured in order to earn an income, but rather as an instrument of self-fulfillment so central to identity that it should not be controlled by others. Therefore, the bishops condemned high unemployment and widespread poverty as a "social and moral scandal," and supported workers' right to unionize, democracy in the workplace, and participation of workers and communities as well as management in deciding on capital allocation and plant closures. Regarding family farms, the bishops declared they "should be preserved and their economic viability protected. . . . As pastors, we cannot remain silent while thousands of farm families caught in the present crisis lose their homes, their land, and their way of life." Likewise, concerning the international arena, the bishops stated that multinational corporations "should be required to adopt a code of conduct encouraging the equitable distribution of their benefits among people in the countries where they operate."[12]

Another indication of the reappraising of ethical principles now under way is the appeal for a "kinder and gentler nation" made by Republican leaders during the 1988 presidential campaign. Although suggestive more of campaign strategizing than of ethical conversion, it is nevertheless significant that Republican managers, apparently judged the compassion theme to be in proper season. "It's clear," declared conservative commentator Kevin Phillips at the time, "that we're looking at the growing Republican realization that the social Darwinist state of conservatism—the dog-eat-dog, survival-of-the-fittest economics—has breathed its last. Even if we don't have a stock market crash, the heyday of uncaring capitalism is over. There's a sense that too many people have been left out and the conservative cycle has passed its peak."[13] The 1989 annual report on federal assistance programs by the House Ways and Means Committee revealed that the average family income of the poorest one-fifth of the U.S. population dropped from $5,439 in 1979 to $5,107 in 1987. In that same period, the average family income of the top fifth increased from $61,917 to $68,775 (all in constant 1987 dollars). Representative Thomas J. Downey commented as follows about these figures: "The invisible hand of the marketplace has got to be tempered by the just hand of Government. You're not going to have enough locks on the doors or police in the streets to protect you from a generation of people who are not part of the mainstream of American economic life."[14]

Perhaps the greatest present-day challenge to capitalism is being raised within its American stronghold in the name of ecological well-being. According to Lester Brown of the Worldwatch Institute, the great advantage of a market economy is its superior efficiency in allocating resources among various possible uses. It avoids the common Soviet dilemma of factories meeting production quotas year after year while their output is piling up in warehouses because the products fail to meet quality standards or changing fashion styles. Market price fluctuations, by reflecting conditions of glut and scarcity, determine quickly and reasonably accurately optimum resource allocation.

Lester Brown points out, however, that the same capitalist marketplace is not geared to register or react to the overexploitation of natural resources before they have deteriorated to the point that prices skyrocket. By then, the affected resource may be beyond retrieval—a loss that has actually occurred repeatedly throughout the globe in recent history. The market mechanism also fails to take account of external costs associated with certain economic activities. Pollution and its resulting acid rain, originating in the smoke-stack industries of the American Midwest and northwestern Europe, are inflicting severe damage on fresh-water fisheries, agricultural croplands, and timber stands thousands of miles away. The producers of the industrial products involved, however, have not been held responsible for the heavy pollution costs, which thus far have been borne by those plant, animal, or human victims who happen to reside on the leeward side of the industrial plants.

In addition, Brown indicates, the market system is preoccupied with immediate bottom-line considerations to the neglect of longer-term human or even planetary perspectives. High agricultural prices, for example, encourage farmers to plant "fence post to fence post" and to ignore soil-conservation practices. In such a situation, only governmental intervention in the form of regulations and subsidies for the needed conservation practices have a chance of persuading the farmer to adopt measures that serve both long-term public interests and his own.

The general problem is that the market mechanism tends to promote private interests far more than public, though these "private interests" may involve multinational corporations with budgets and resources larger than those of many Third World countries. To cope with this dilemma, Lester Brown suggests that the time has come to reconsider the validity of the current growth Olympics in which all governments are engaged. The size and annual growth rate of a country's gross national product have become criteria of national prestige and well-being. But much of the economic growth now underway is in the long run impoverishing, not enriching, because of the ecological damages that ensue. Just as an economic "deflator" is commonly used to factor inflation into growth statistics, so an ecological deflator is needed to account for the erosion of national or global resource bases. Use of such an ecological deflator would make possible, for instance, a comparison of sustainable food production with projected growth in world food demand; or conversely, it might allow for the definition of what proportion of current world food production is based on the unsustainable use of natural resources. Such ecological concepts and statistical calculations need to be developed systematically, and to be factored in when national and international development projects are prepared.

From a socioeconomic viewpoint, while granting the unprecedented productivity of the global economy operating mostly on capitalist principles, it is obvious that the popular axiom about the rising tide floating all ships is not operative today. Yachts admittedly are riding high, but many small craft remain stranded, whether they be Third World countries now enduring declining per capita incomes, or a growing underclass living in increasingly Third World–like conditions in affluent First World countries. Educator and business executive Clifton R. Wharton, Jr., has observed that, following World War II, Western leaders assumed they had the solution for the problems of the newly independent countries of the Third World. "Now, we find to our shame that not only did *we* not know all the solutions to *their* problems, we do not even know the solutions to our own—and that *our* problems and *their* problems may be, if not identical, of more than passing resemblance." Wharton concludes that "we have the disturbing evidence that we are in fact becoming a two-tier nation in a two-tier world, with the same potential for explosive conflict between the haves and have-nots within our own borders that once alarmed us elsewhere around the globe."[15]

Symptoms of Third World conditions within the United States have become commonplace in newspaper reports and on television screens: 20 million Americans who are chronically hungry; 13.5 percent of the population reported by the Census Bureau as living in poverty in 1987; hospitals throughout the country closing down their emergency centers because too many patients lack medical insurance or other means of paying; and the incidence of homelessness reaching epidemic proportions.

Why should social afflictions on such a scale prevail in the foremost capitalist country of the world? Some shrug it off as an ephemeral and superficial disorder. Joseph Nye, director of the Center for Science and International Affairs at Harvard's John F. Kennedy School of Government, and close adviser to Governor Michael Dukakis, has made a sharp distinction between American and Soviet economic problems. The American ones he considers to be "relatively short-term—the folly of not paying our bills in the 1980s." By contrast, he holds that "Soviet problems go deep into the very nature of their economic system."[16]

Nobody, least of all the Soviets themselves, would deny the gravity of their societal problems. But many would take issue with Nye regarding the "short-term" nature of United States problems, which actually go back at least half a century to the Great Depression, a time of economic prostration and psychological trauma for global capitalism. The trauma persisted until World War II provided an insatiable market for previously closed or underutilized factories, and ample jobs for the unemployed.

With the end of the war, both political and business leaders were apprehensive about returning to the freewheeling economic style of prewar capitalism. An alternative was offered by the novel theories of John Maynard Keynes. Whereas Karl Marx had been the theorist of class conflict between capitalists and proletarians, Keynes was the theorist of class compromise. His diagnosis and his prescription were appealingly simple and persuasive. If workers cannot afford to purchase with their wages what they produce in the factories, the inescapable end result is a ruinous cycle of inadequate purchasing power, business slowdown, wage cuts, layoffs, closed factories, and growing unemployment until finally a new and less favorable equilibrium is reached. Keynes's remedy was for the government to intervene with various measures designed to increase mass purchasing power and block or reverse any downward cycle. The singular appeal of this prescription was that it seemed to suit the interests of both workers and their employers.

During the prosperous post–World War II years, Western governments painlessly implemented Keynes's basic strategy of "priming the pump" and creating "effective demand" with such measures as progressive taxation, trade-union recognition, and social legislation of all sorts. Keynes's historic class compromise contributed decisively to decades of unprecedented economic growth and social development. Those were the halcyon decades of the welfare state—limited though it was—which was the institutional instrument for implementing the combination of measures comprising the Keynesian kit.

The Keynesian compromise can work only within the parameter of national boundaries. These boundaries have in recent years become increasingly porous because of advances in communications, transportation, and management, which allow corporate entities of almost unimaginable size to locate and relocate their operations freely throughout the world. Individual corporations have reaped immediate profits from this new international balance, but while doing so they have engendered serious long-term problems for global capitalism. When an auto company closes its Detroit plant and opens a new one in Mexico where it pays its workers two dollars an hour as against twenty dollars at home, the result is a sharp drop in auto sales in both countries. Neither the unemployed Detroit workers nor the underpaid Mexican workers can afford to buy new cars. The outcome is an economic downward spiral which may one day precipitate a new Great Depression. Worldwide lack of purchasing power generates pressure to lower wages in First World countries and to keep them low in the Third World. This threatens to resurrect the dilemma of overproduction (or underconsumption due to weak purchasing power) that was the hallmark of the Depression years half a century ago. Symptomatic of this ominous trend is the shrinking of world trade growth from average annual rates of 8 percent in the 1950s and 1960s to less than 3 percent in the 1980s.

Some economists argue that one way out is to restore the Keynesian historic compromise, which, given today's integrated global economy, must take place on a worldwide scale to be effective. Either Third World wage scales must be raised, or First World wage scales will be forced down. Various proposals have been made for reducing the current gap between the two, such as making Third World access to First World markets contingent upon meeting international standards regarding wage levels, working conditions, and social benefits. More specifically, it has been proposed that imports from cheap-labor countries be subjected to a "worker exploitation tax" equal to the difference in wages paid to American and foreign workers. The revenues produced by this tax would then be transferred to the workers in the particular Third World country in question in order to raise its prevailing general wage level. The difficulty in reaching such a global arrangement and actually carrying it out can be imagined by recalling how bitterly congressional and business interests opposed proposals to raise the American minimum hourly wage above $3.35.

The strategy of global Keynesianism has been directly challenged by conservative economists who maintain that the current difficulties of capitalist countries such as the United States stem from excessive rather than insufficient purchasing power. The excess, they believe, results in a growing demand for imports and a skyrocketing trade deficit. Their remedy is to reduce consumer demand by decreasing government-spending in entitlement programs, and to increase business profits by reducing wages, speeding up the pace of work, and encouraging technological innovation. Prime Minister Margaret Thatcher, who has enthusiastically implemented these policies, has proclaimed that they are in accord with the Sermon on the Mount. In May 1988 before the General Assembly of the Church of Scotland, she also invoked St. Paul—"If a man will not work, he shall not eat"—to argue that the rich were blessed while the poor were not. "Indeed, abundance rather than poverty has a legitimacy which derives from the very nature of Creation." The implications of this brand of economics and morality are indicated by a 1988 Harris poll which found that a majority of British citizens thought their country had become richer and freer of government in the last ten years, but by a four-to-one majority they also thought the country had become more selfish, and two out of three thought it was more unhappy.[17]

The future of these and other conflicting capitalist strategies is likely to be determined by the nature of the "real world" in which they will be debated, a real world in which neither global Keynesians nor Thatcherites of any stripe may prove to be "winners." If there is a "solution" to the present global crisis, it may lie in economic and social thinking not yet apparent to many—or perhaps not yet in existence. That a crisis of major proportions is under way can, however, hardly be doubted. In December 1987, thirty-three economists from thirteen countries assembled in Washington by the Institute for International Economics warned that "something is seriously wrong with the world economy." They noted that the bond market had fallen by 30 percent in the early months of 1987, and the global stock markets by 20 to 30 percent in the fall of the same year. "A third crash of the markets," cautioned the economists, "could be greater than either predecessor, with far more pervasive results."[18]

What these "pervasive results" might be were not spelled out by the assembled economists, nor could those economists be expected to come forth with specifics. Individual countries have individual historical and cultural traditions that result in distinctive institutional outcomes and performances. Nineteenth-century capitalism assumed a dominant form in Britain that was quite different from the variants that developed contemporaneously on the European continent and in the United States. Twentieth-century capitalism has been dominated for the most part by the United States, yet very different and significant variants are taking form today in other regions, such as Japan and Scandinavia. Even within the limited confines of western Europe, a struggle is under way with the approach of 1992, the year when that region discards trade and other barriers and forms a new European Community or Common Market. The issue is what type of capitalism should prevail in the Community. Should it be a free-enterprise economy dictated largely by the market (as favored by Thatcher's Britain, backed by the United States and the multinationals), or a managerial economy with a welfare state and worker participation in decision-making (as favored by Germany, which has the largest economy in western Europe)?

It may be assumed that in the forthcoming decades this regional diversity will persist and intensify as capitalism takes root in lands with infinitely more varied historical and cultural backgrounds, and as the technological precociousness of capitalism continues to make "creative destruction" the mighty social sledgehammer of our times. The capitalist world of the late twentieth century is a world groping toward alternative capitalisms, capitalisms that seek to reconcile what the *Forbes* editors dismiss as irreconcilable—"to protect the victims of change" (whether human or ecological) without stifling capitalism's creative impulse for change and for growth. If the groping for alternative capitalisms appears feeble at present, it should be recalled that during the 1930s the unorthodox theories of Keynes quickly became orthodox in response to the trauma of the Great Depression. Given the unprecedented power and unpredictability of today's capitalist sledgehammer, it appears unrealistic to dismiss

207

the possibility of new traumas spawning new orthodoxies with new institutions and practices—a polymorphic capitalist world compared to that of the nineteenth and twentieth centuries.

TOWARD ALTERNATIVE SOCIALISMS

The search for alternatives is under way in the socialist as well as the capitalist world, and the leaders of the socialist world are very much aware of what is at stake. In January 1987, Gorbachev told his party's Central Committee that "the development of microelectronics, computer equipment, instrument building, and the whole information industry is the catalyst of present-day scientific and technological progress. . . . a new stage of the scientific-technological revolution ensuring a manifold increase in labor productivity, huge savings of resources, and improvements in the quality of output, is beginning. . . . [Soviet economic progress depends on] how skillfully we are combining the advantages of socialism with the achievements of the scientific-technological revolution."[19]

Gorbachev and other socialist leaders are particularly sensitive to the need for adapting their societies to the imperatives of high technology exactly because they have thus far failed to do so, with painful consequences that include inefficient management, poorly motivated workers, substandard consumer goods, and services in short supply. Today the socialist world with its Marxist model of social reorganization is experiencing far more intense soul-searching than the capitalist world with its model of creative destruction.

If socialism is defined as an ideology stressing cooperation and collective well-being rather than private profit and individual acquisitiveness, then its roots go back many millennia to the time when relatively egalitarian Paleolithic bands gave way to agriculturally based class-stratified civilizations with a few "haves" and many "have-nots." The resulting inequity and exploitation led countless reformers and prophets through the ages to suggest plans for promoting social justice. In the classical world, for example, Plato in his *Republic* called for an aristocratic communism, a dictatorship of communistic philosophers. In medieval times, the English priest John Ball prescribed a drastic remedy: "Good folk, things cannot go well in England nor ever shall until all things are in common and there is neither villain nor noble, but all of us are of one condition."[20]

This egalitarian impulse, which usually took a religious form in the various civilizations, found secular expression in modern times. The turmoil and passions of the English and French revolutions, and of the contemporaneous Industrial Revolution, engendered many schemes for social reform. First came the utopian socialists (Claude-Henri de Saint-Simon, Charles Fourier, and Robert Owen) of the late eighteenth and early nineteenth centuries, followed by Karl Marx (1818–83) and his "scientific socialism." Scientific or not, socialism remained but a theory and a dream until the 1917 Bolshevik revolution and the founding of the Union of Soviet Socialist Republics. During the 1930s, the juxtaposition of the ambitious Five-Year Plans and of the capitalist Great Depression roused high hopes for the Soviet socialist model. These hopes have since been dimmed both by the lack of individual freedoms in the USSR, and by the failure of Soviet technology and the Soviet economy to keep up with its capitalist competitors, whether in Japan, western Europe or the United States.

Not only has the USSR failed as a socialist model—whether judged economically or in terms of gains in social freedom—but so have the numerous other socialist societies that emerged in Europe, Asia, and Africa following World War II. If the destructive side of capitalism has created immense problems for the world, socialist attempts to restructure societies have spawned even more serious and immediate problems in countries like China, Vietnam, Yugoslavia, and Ethiopia, as well as in the Soviet Union itself. Two general factors stand out as primarily responsible for the difficulties experienced by these widely scattered socialist societies.

The first is their uniformly underdeveloped and impoverished historical origins. Marx had assumed that his long-awaited revolution would occur in the developed industrialized countries before it did in their colonies. Instead, post–World War II revolutionary socialism emerged mostly in the underdeveloped, poverty-stricken former colonial or semicolonial areas. Socialism everywhere (with the partial exception of eastern Europe) has appeared on the historical stage as a substitute for rather than a successor to capitalism.[21]

This is an immensely significant historical fact, for it means that all socialist societies were born with a built-in and inescapable dilemma.

They could have concentrated on overcoming their inherited poverty (often made worse by devastating warfare preceding national independence) through competitive individualism and market processes, which would have generated the very inequity and injustice that socialism is designed to eradicate, without necessarily guaranteeing economic success. Or they could have focused on developing egalitarian socialist societies with corresponding socialist institutions, a process that has usually slowed economic growth and perpetuated inherited poverty and backwardness—a politically vulnerable course in a world of accelerating and infectious consumerism. The cruel dilemma imposed on the socialist world by history has forced socialist societies either to give primacy to economic development and relegate socialism to a receding future, or to focus on socialist objectives that ensure poverty and backwardness for the foreseeable future. The difficulty of holding consistently to either course has led many socialist states to vacillate in a damaging way between programs emphasizing productivity first or equity first, as exemplified in China's recent shift from Mao's "Serve the People" to Deng's "Enrich Yourself" policies.

As harmful as the uncertainty over equity versus productivity has been in the economic field, ambivalence in the political field between dictation from above and mass participation from below has been even more crippling. This political ambivalence also is an inherited one—both from the recent past in which Marxist-Leninist ideology developed, and from a more distant past in which the values of tributary society held sway. Marxist-Leninist ideology has from its origins viewed the Communist Party as both servant and master of the people. The party, as a result, has been expected to tread an exceedingly thin line between "tailism," a passive following of mass desires, and "commandism," an elitist exercise of dictatorial power over the masses.

In addition to the ideological problem, there is the historical fact that socialism has emerged successfully only in countries with virtually no democratic traditions. China, to take an example, has been described as a country "where two virulent bureaucratic traditions ..., that of the Chinese mandarin and the Communist *apparatchik*, make their influence felt simultaneously."[22] When the Maoist regime finally was established on the mainland in 1949, it was the outcome of devastating decades of warfare between the Red Army, the Nationalist army of Chiang Kai-shek, and until 1945 the Japanese forces occupying large parts of the country. Likewise, in eastern Europe, dictatorial regimes prevailed everywhere when World War II began, with the solitary exception of the Czechoslovak republic. In Russia the watchwords of the tsarist autocracy had been "orthodoxy, autocracy, nationalism," which summarize neatly the essentials of the Stalinist autocracy.

It is in the light of this historical background of economic underdevelopment and political authoritarianism that current problems in the socialist Second World should be interpreted. Russia's Mikhail Gorbachev has been notably forthright in emphasizing the need for a break with such past institutions and practices. He has gone much further than his reformist predecessor, Nikita Khrushchev, who tried to deal with Stalinism by using the analytically limited concept of the "cult of personality" to explain the decades-long failure of a system.

Gorbachev, by contrast, has proclaimed publicly and repeatedly that a reappraisal of fundamentals is essential. In June 1987, for instance, before the Communist Party Central Committee he declared: "Ours is a rapidly changing society . . . a society with new attitudes and new hopes. . . . We are in for new problems, considerable complexities. We are not insured against mistakes. . . . Yet I am confident that the greatest mistake is fearing to err."[23]

Gorbachev's phrase, "fear to err," is strikingly reminiscent of President Roosevelt's phrase, "nothing to fear but fear itself." In fact, a revealing parallel may be drawn between the New Deal of the 1930s and the *perestroika* of the 1980s. Both programs were designed in response to a traumatic crisis—Hoover's Great Depression in the United States, and Brezhnev's great stagnation in the Soviet Union. Both programs were headed by leaders who came not from below but from above—from the national elite. Both Roosevelt and Gorbachev were interested not in a revolution that would overturn their social systems, but in reform substantial enough to rejuvenate and preserve those systems. "Our society will never again be what it was," declared Gorbachev in May 1988. "It is changing. . . . A great deal is to be done, but the train has already started off, and it is gathering speed." At the same time, the Soviet political scientist and Gorbachev adviser Fyodor M. Burlatsky declared, "An immense struggle is unfolding within Soviet society. It is taking place on the basis of socialism, but the notions we have had of socialism are changing. . . . All is in flux, and the future is at stake."[24]

The prospect of such sweeping change understandably aroused strong opposition from conservatives who were fearful of an uncertain future, and by radicals who demanded societal surgery rather than first aid. Thus, FDR was excoriated from the right as a traitor to his class, and from the left by adherents of Huey Long's Share-Our-Wealth Society. The ferment in the Soviet Union today is just as yeasty. The younger generation, now assuming positions of power, grew up during the Brezhnevian years of stability. It is more educated, more aware of the shortcomings of the Brezhnev regime, more critical of the present, and more demanding about the future. In addition, there are conservative elements, national minorities, technocratic interest groups, and reformist circles of varied political hues. Consequently, the process of implementing Gorbachev's *perestroika* has required political juggling at least as delicate and complex as that which faced the New Dealers in Washington a half century earlier.

To his right, Gorbachev faces the Russian nationalists, who favor a traditional planned economy geared toward military and heavy-industrial priorities. In political and cultural matters, the conservatives oppose the pluralism and nonconformity implicit in *glasnost*. Typical is novelist Yuri Bondarev, who complains that the newly unleashed press "destroys and denigrates," and greets hallowed concepts like "patriotism" and "motherland" with "serpentine hissing." Bondarev castigates the supporters of *glasnost* as "civilized barbarians" intent on destroying native Russian culture, and he goes so far as to call for a new purifying Stalingrad battle in which aroused patriots would repulse the insidious barbarians from within. Some who share Bondavrev's views belong to the organization Pamyat (Memory), which is strongly anti-Semitic and anti-Western. It views Gorbachev and his supporters as excessively Western-influenced, and advocates a strong autocracy in "the Russian tradition."

Restive minorities also present a formidable and growing problem for Gorbachev. One reason is that tsarist Russia and the Soviet Union never became a "melting pot" as did the United States. Whereas the United States naturally took the "melting pot" route as millions of immigrants from all over the world streamed into relatively "empty" lands, the tsarist empire was formed by imperial armies conquering numerous ethnic groups, which then continued to exist as discrete components of an imperial mosaic rather than being fused into one relatively homogeneous mass. That is why the Union of Soviet Socialist Republics is a union of sixteen republics that are officially Soviet and socialist in nature but which, more importantly, are all based on ethnic foundations—Russian, Ukrainian, Azerbaijan, Uzbek, Tadzhik, Lithuanian, Latvian, Estonian, Georgian, Armenian, and so forth.

These distinct and enduring ethnic groups comprise latent opposition elements that can be mobilized and activized whenever divisive issues develop within Soviet society. One such issue is the fear of Russification, aroused as Russian workers migrate to Baltic, Central Asian, and other republics to work in newly built factories. In the Latvian Republic, for example, the Latvians now find themselves a minority. In October 1988, responding to this and other issues, Latvians formed the Latvian Popular Front, which has refrained from demanding full independence for Latvia but is pressing for virtually all other attributes of independent statehood, including the right to create its own currency, the right to establish diplomatic relations with foreign countries, and the right to limit the immigration of Russians and other Soviet citizens to the Latvian Republic. "For over 40 years," declares a prominent Latvian nationalist, "I have watched the culture and the economy of my country slowly deteriorate. The time has come for us to take back control of our own land because the loss of a true Latvia is no longer a threat; it is a real and pressing danger."[25]

Fear of Russification is not the only concern of restive Soviet nationalities. Armenians and Azerbaijani are at loggerheads because the historic intermixture of these nationalities going back to tsarist times has left minorities of each within the other's republic. Hence, the growing pressure on Moscow to redraw republic frontiers to conform more closely with present-day ethnic realities. Some ethnic leaders who oppose Gorbachev's *perestroika* for reasons having nothing to do with ethnic considerations are also not above exploiting ethnic consciousness as a convenient tool with which to resist reform.[26]

Whatever the root cause, Soviet minorities represent a serious challenge to Gorbachev's program. Thus far, what they have manifested has been closer to ethnic self-assertiveness than anything approaching the anticolonial drives of Third World peoples who were determined to win full independence and who refused to settle for anything less. One task confronting Moscow policy-makers is to prevent current demands for more freedom within the USSR from escalating into demands for freedom from the USSR. Nationalist discontent and agitation are escalating so rapidly within the republics that it is becoming increasingly questionable whether the fine line separating autonomy from independence will be respected and preserved.

Another important force in the current Soviet maelstrom consists of technocrats who are generally supportive of the *perestroika* concept. Comprising highly trained economists, scientists, academicians, journalists, and other professionals, they publish at home and abroad sophisticated articles criticizing the existing "command-administrative system" and calling for a more efficient type of "market socialism." More specifically, they tend to favor ending food subsidies, requiring industrial enterprises to be self-financing even at the cost of plant closings and unemployment, legalizing small private businesses, and encouraging hitherto disparaged desires for personal enrichment and conspicuous consumption. Nikolai Shmelev, a prominent economist and Gorbachev adviser, aptly summarized such technocratic sentiments when he advised his fellow countrymen not to be afraid of losing their "ideological virginity."[27]

Soviet technocrats are comparatively well known abroad, where their promarket prescriptions are understandably welcomed. Less well known is another element somewhat related to the technocrats and yet profoundly different. This group consists of leftist intelligentsia, sometimes referred to as the Soviet New Left. They also support *perestroika*, but insist that it can prevail only as a movement from below rather than as a program imposed from above. Therefore, they prefer to identify themselves with the word *obnovlenie*, or democratic socialist renewal. This word has certain connotations that distinguish Soviet leftists from Soviet technocrats.

For many of these leftists, the attainment of spiritual goals rates as high as raising the GNP; blind worship of the marketplace is not considered a significant advance over worship of a central plan; acceptance of mindless consumerism is rejected as firmly as acceptance of bureaucratic authority; and more free time is valued as highly as more private income. In short, Soviet radicals want first and foremost to place the satisfaction of basic human needs on the *perestroika* agenda.

In recent years, informal groups espousing this radical agenda have sprouted throughout the country "like mushrooms after rain," to use a Russian expression. Representatives of these groups who met at a conference of "Socialist Civic Clubs" in Moscow in August 1987 declared their basic aim to be the completion of the democratic socialist revolution begun in 1917 but "derailed" by the triumph of Stalinism. They expressed support for Gorbachev's "market socialism," but only on condition that it be coupled with "firm guarantees for the preservation of the social conquests of the working people—full employment, minimum wage, pension security, etc." One of the conference participants declared that since their aim is socialism based on self-management, they should study the experience of the U.S. New Left of the 1960s, with its "ethic of cooperation and repudiation of predatory competition and individualistic pursuit of egoistic personal interest."[28]

The reference to the American New Left is significant because like them the Soviet left intelligentsia represents only a small minority of its country's people. This raises the crucial question of the views of Soviet workers and peasants who comprise the overwhelming majority of the total population and who appear to be deeply ambivalent about developments, perhaps because their status is, in fact, highly equivocal. On the one hand, they have plenty of firsthand knowledge of the corruption and cronyism that characterized the preceding Brezhnev regime. On the other hand, they have enjoyed a secure and comfortable niche within that regime, including guaranteed jobs, as well as food, housing, and medical care that have been subsidized and cheap, even though of mediocre or poor quality. The Soviet people as a whole have learned to accept their socialist society as a national insurance policy, whose security is generally welcomed and popular. What is not popular is the quality of the goods and services provided, especially as more information is received on the high-quality goods and services available in the West. Soviet audiences watching American movies like *Kramer vs. Kramer* usually are deeply impressed not by the quality of the acting but by the size and furnishings of the Kramer apartments pictured in the background. The ideal arrangement for most Soviet citizens would involve an improvement in the quality of goods and services together with a continuation of the existing lifetime security policies. They are shrewd enough, however, to realize that such a combination is highly uncertain, if not unlikely, in the foreseeable *perestroika* future.

So the Soviet citizenry, while they may yearn for those birds in the bush quite reasonably hesitate to let go their grip on the bird already in their hand. Such an ambivalent stance poses a formidable obstacle for Gorbachev. His *perestroika* (with its market socialism and its threat to close inefficient plants and fire underproductive workers) is, in effect, an offer of a Soviet-type of New Deal—less economic security in return for a fatter paycheck for those who pass muster. Gorbachev has tried to blunt the sharp edges of this harsh quid pro quo by stressing *glasnost.* If Soviet citizens really become involved, and if they really contribute their energy and their creativity, he claims that then they will not have to choose between security and higher living standards. A united and purposeful national leap forward would enable them to have their cake and eat it, too, leaving Soviet citizens better off than their Western counterparts who continue to bear the heavy burden of insecurity in order to enjoy the consumerism so envied throughout the Second World.

Roosevelt's New Deal never had to undergo a full test of its viability, thanks to the "Good War" that opened factories and created jobs. Gorbachev's New Deal can expect to find no such escape valve in today's nuclear world. So the USSR of the 1980s resembles in certain ways the USA of the mid-1930s—the same unrest, anxiety, dreams, nightmares, and impatient social experimentation that is at once exhilarating and sobering. The eminent sociologist Tatiana Zaslavskaia has repeatedly emphasized that many years of theorizing, strategizing, and listening to Soviet citizens will be necessary before a broadly acceptable and workable set of policies has a chance of emerging. The first steps in this process include a new attentiveness to the experiences of other countries, as Gorbachev's chief economic adviser, Abel Aganbegyan, informs us. "We are learning from experience in Eastern Europe and in China and also from capitalist countries. . . . some elements of each may be useful and we shall draw on them."[29] Likewise, economist Nikolai Shmelev has stated that "it may be necessary to fire 1% or 2% of the people in an enterprise who are drunkards or just don't work." But he added that 10 percent unemployment, which is not uncommon in the West, could never be tolerated in the Soviet Union. "I don't want any capitalism in our country. I don't want to lose our social net. . . . I feel that you Americans are some kind of extreme and we Soviets another, and somewhere in between is the truth."[30]

Meanwhile, Gorbachev, as Soviet political leader, is attempting to dissipate mass skepticism and mobilize active support for his program. Typical is his effort to win over the peasants by offering them fifty-year leases on private plots in place of the current practice of working on collective farms and collecting a government salary regardless of the collective's productivity. Collective farming must end, Gorbachev has argued, because "no fool is going to go to work on a lease contract as long as he can have a salary without earning it."[31] Despite Gorbachev's logic, opposition to his private land leasing was quickly expressed by collective farm bosses reluctant to give up their positions, by farmers preferring the collective's security to the uncertainty of private farming, and by the general public which, after decades of egalitarian indoctrination, resents the prospect of a new wealthy entrepreneurial class arising in the countryside.

Alexander N. Yakovlev, a principal theoretician of *perestroika* and a member of the Communist Party's ruling Politburo, has stated candidly that the deeply rooted "conservatism" of the Soviet people is a major obstacle to Gorbachev's reforms. At a meeting of party leaders, he conceded that "while we talk about political transformations, how difficult it is to part with the old power to which we are accustomed." As for the mass of the Soviet public, he admitted that they are deeply disturbed by the new principles. "Democracy, openness, cooperative ventures, self-financing, the socialist market, self-management, the sovereignty of the people, the pluralism of opinions—all this is penetrating our life deeply. It disturbs our life and sometimes deprives certain people of sleep and rest. . . . people are nostalgic for the old certainties."[32] How many people in how many different societies are experiencing this nostalgia in our age of unprecedented, ever-accelerating, and extremely unsettling change!

TOWARD THIRD WORLD ALTERNATIVES

If the First World is reluctantly beginning to consider the possibility of alternative, more life-enhancing capitalisms, and the second is actively groping toward alternative decentralized and more productive socialisms, the Third World is in an even worse predicament. It is not so much groping toward anything as floundering. Vulnerable and constantly manipulated by outside interests on behalf of outside interests, it lacks a sense of direction, agreed-upon goals, and agreed-upon strategies for reaching any goal whatsoever.

The wavering career of Ghanaian leader Lieutenant Jerry Rawlings reflects exactly this predicament. When he first seized power in 1983, he proclaimed a new socialist-oriented Ghana and turned for aid to Libya and the Soviet Union. When that aid began to dry up and the Ghanaian economy with it, he turned to the West for help. He followed the injunctions of the International Monetary Fund—cutting social spending and instituting "austerity" measures—so faithfully that the IMF dubbed him "a role model for all of Africa." The Jerry Rawlings interviewed on the CBS News program "60 Minutes" in April 1988, proved to be a very different man from the swaggering flight lieutenant who several years earlier had overthrown a civilian government in a coup lasting only a few hours. Superficially, Ghana's shift to the West appeared to have been a winning strategy. Bustling city markets were crammed with goods, the gold mines were operating full blast, and cacao production had doubled. Yet Ghana's rising economic tide, admitted Rawlings, had not "lifted many boats. . . . it's a painful thing for me, sitting here, to have to be making such admissions. . . . People still don't have money in their pockets to buy the things that are now in the stores. . . . you've seen so many hundreds of thousands of people dying away from deprivation. And I'm sitting here, there's nothing I can do to help them. . . . it's like clawing in the dark, looking for a way of dealing with some of the global problems that have reduced Africa to this kind of situation. I don't want to be pushed to venting my feelings. . . . It may not be politically right to go too far, but its painful. . . . Sitting so helpless. So helpless. . . ."[33]

In the immediate post–World War II years, Third World countries did not seem to be "clawing in the dark." Instead, they set about purposefully and single-mindedly winning independence from the European colonial powers, a goal they reached with unexpected speed during the 1950s and 1960s. The obvious next goal was to win economic and cultural independence commensurate with the political. At this point, the floundering set in, and it has persisted to the present day. It persists with both types of Third World states that emerged from the independence struggles: the conservative nationalist states and the social revolutionary ones.

The leadership of the conservative nationalist states were mostly Westernized merchants, teachers, clerks, officials, and military officers. They were the sort of people who wanted primarily to end foreign rule but not to shake too drastically already existing social institutions. They were not Marxists, nor did they espouse class conflict. They did not challenge in a fundamental way local or foreign vested interests, whether traditional landholdings, plantations, commercial firms, banks, railways, mines, or government debt arrangements. Such nationalist leaders were more likely in the end to be entrusted with political power by their former colonial masters because it was understood implicitly that they would not use their power to effect basic social or economic changes. Britain, for instance, granted independence to an India led by the relatively conservative Gandhi, Nehru, and Congress Party, but fought the Malayan Communist guerrillas to the bitter end. France granted independence to the conservative Francophile Félix Houphouët-Boigny in the Ivory Coast, but fought the Communist Ho Chi Minh in Vietnam and the socialist Mohammed Ben Bella in Algeria. Likewise, the United States gave independence to the Philippines but waged full-scale war against Ho Chi Minh for another two decades to ensure that Vietnam would not end up as a "fallen domino." Similar strategic calculating in Africa led the British to grant independence to nationalist leaders in Ghana and Nigeria, while the Portuguese fought to the end against the Communist-led movements in Angola and Mozambique.

After attaining independence, these nationalist states set out to spur the economic growth to which they all aspired, but within the context of the global capitalist order. They increased as much as possible the production and export of raw materials; they negotiated loans to finance their growth; and they accepted development plans and technical aid from Western governments and international bodies such as the United Nations, which sponsored two "Development Decades." With these policies, the nationalist leaders hoped to replicate the process of economic growth and eventual full industrialization pioneered in western Europe and the United States in earlier centuries. Such a replication has, except in a few isolated cases, proved impossible. The global economic setting has changed too fundamentally for underdeveloped states to repeat the process of industrialization of past underdeveloped states.

The historical experience of England provides a good example of the difference between past and present industrialization. During the early stages of the Industrial Revolution, English workers were severely exploited because medieval guilds had atrophied and trade unions were not yet legalized. Hence, the machinery-smashing by Luddites and the 1819 "Peterloo massacre" of rioting workers. After 1850 the "trickle down," so talked about in the Third World today but which so rarely materializes, did actually occur and did benefit the British working class, resulting in industrial peace and general prosperity. But this was possible only because British industry enjoyed the unique advantage of a monopoly of world markets. By 1814 Britain was exporting 14 percent more cotton than was being used at home, and by 1850 the differential had increased to 24 percent. Other industries also benefited from the access to global markets, thus enabling Britain to become the "industrial workshop of the world."

Developing countries today have no such broad avenue beckoning to the future—no such foreign markets open to their products. Instead, world price levels for their raw materials continue steadily to decline, while their few industrial exports (textiles, steel, shoes, and clothing) encounter increasing demands from the already industrialized world for "fair trade"—the most recent euphemism for protectionism. Equally serious for Third World countries aspiring to climb up the economic ladder has been the inelasticity of their own domestic markets. Native industries not only have problems of access to foreign markets but also are fettered by inadequate purchasing power at home, the two phenomena being, of course, linked.

220

Underpaid industrial workers and displaced peasants have experienced little of the "trickle-down" effect seen in western Europe, the United States, and Japan in their earlier industrializing experience, and presumed by many economic theorists to be part of the natural "takeoff" process. Instead, there has been "trickle up" to local elites, and "flow out" to multinational corporations that control international trade, and to foreign governments and banks that receive the annual payments due on the enormous Third World debt, which by 1989 totaled $1.3 trillion.

Financial relations between First and Third World countries have been managed by the First World partially through the International Monetary Fund (IMF), which requires debtor nations to accept its structural adjustment programs (SAPs) in return for an IMF "stamp of approval" that is a prerequisite for credits and investments from First World sources. The main thrust of SAPs is to encourage or force the sale of government-owned enterprises to private interests; the construction of harbors, railways, and roads to facilitate the export of raw materials; the ending of state subsidies for basic staples that make food affordable for the poor; and the reduction of expenditures for health care, education, and social services in order to leave sufficient funds to meet debt obligations. This "demand management," to use IMF terminology, has proven eminently successful—just not for the Third World. According to figures from the Organization for Economic Cooperation and Development, between 1982 and 1987 the Third World as a whole received development assistance, export credits, bank loans, and private investments totaling $552 billion. During the same period, Third World nations paid out $839 billion in interest and amortization on their debts. Thus, the poorest nations on earth provided the most affluent ones with a net gain of $287 billion in a mere six years—a sum approximately equal to four Marshall Plans. Even after paying out this amount, the poor nations were one-third deeper in debt in 1987 than they had been in 1982. Inevitably, such hemorrhaging affects local living standards. Between 1980 and 1987, Latin America's per capita income dropped 30 percent and in sub-Saharan Africa the decline has been even greater.

Such economic facts of modern life have had profound repercussions politically, helping to push Third World societies toward ever-greater authoritarianism. Beset by economic difficulties, beleaguered by First World demands, governments have been unable to satisfy their newly aroused publics. Hence, the chronic conflict between labor organizations on the one hand, and on the other the governments trying to enforce austerity measures necessitated by outside pressures for "demand management." In country after country, this conflict culminated in recent decades in coups led by members of civil or military bureaucracies. Lacking a strong and independent economic foundation, Third World bureaucracies often depend entirely on the armed forces and state mechanisms for their power and their sustenance. Therefore, they are forced to resort to any means to cling to the official posts that shield them from the world of poverty that surrounds them. This helps explain the paradoxical Third World pattern of external colonial domination being overthrown, only to be replaced by local military dictatorship or one-party rule dependent on the largesse of their former First World masters.

Negative economic statistics had social as well as political repercussions, involving as they did drastically reduced social services and correspondingly increased malnutrition and illness. An estimated 15 to 20 million people die each year in the Third World from starvation or malnutrition-related diseases. This is the equivalent of a Hiroshima every two days. The ultimate irony is that the global imbalance resulting in a net outflow of capital and even of foodstuffs from the poverty-stricken South to the affluent North hurts the North as well as the South. Because debtor countries cannot both service their debts and pay for imports, U.S. exports to Latin America and Africa dropped from $52 billion in 1981 to $40 billion in 1987. Every billion dollars in lost exports equals 24,000 lost jobs, so the cost of Third World debt to the American economy was 288,000 jobs in 1987.[34] A United Nations economist has analyzed these contradictions in the global economy as follows: "In contrast to the Depression of the 1930's, today there is genuine worldwide economic interdependence. Third World debt problems and low export earnings are forcing developing countries to close their markets to imports from industrialized nations. The recessions are feeding on each other. The collapse of the Third World will hurt the First World's recovery."[35]

U.S. Secretary of State George P. Shultz responded to this situation in August 1988 by declaring in Rio de Janeiro that Latin

American countries could make economic progress only through "resolute implementation of outward-looking policies aimed at trade and exchange liberalization, deregulation, privatization, and market-based principles."[36] Some economists, however, have raised important questions about the discrepancy between the theoretical benefits of "market-based principles" and the actual end results in the Third World. Take the distinguished Argentine economist Raúl Prebisch, who, following World War II, pioneered an import-substitution strategy designed to stimulate Third World industries to produce manufactured goods hitherto bought from First World countries. The strategy was widely adopted, yet by 1979 Prebisch was acknowledging that he had erred in seeking an economic solution to a problem that he now believed was primarily sociopolitical. "We thought that an acceleration of the rate of growth would solve all [Third World] problems. . . . This was our great mistake." Prebisch noted countries with thirty years of high GNP growth, but growth from which 40 percent of their populations received no benefits. The reason, he explained, was that "we are resisting changes in the social structure," so that the beneficiary of economic growth is only a miniscule and unproductive "privileged consumption society." Any development plan, he concluded, is bound to fail without "a complete social transformation."[37]

Prebisch's prescription for "complete social transformation" was not new. Indeed, all the post–World War II social revolutionary states were based on precisely this principle. Whereas the nationalist states had been concerned primarily with ending colonial rule, the social revolutionary states searched for new social as well as political orders—a search that has proven as ineffectual as those of the nationalist states.

The world's pioneer revolutionary state was the product of the 1917 Russian Revolution. The Bolsheviks were the first ruling group to reject the Western capitalist model, and to set out to fashion a new socialist society at home, while encouraging other revolutions throughout the world. However, with the failure of a post–World War I surge of revolutionary movements in central and eastern Europe, the Bolshevik regime found itself an isolated and economically primitive socialist island in a capitalist ocean. That isolation ended only when the Second World War let loose a new wave of revolutionary activity, beginning in China and spreading to Southeast Asia, Portuguese Africa, and Cuba. Communist regimes were also established in eastern Europe, two—Yugoslavia and Albania—gaining power by their own efforts, and the rest being installed by advancing Soviet armies.

The new revolutionary states scattered about the globe faced the same problems as the original Soviet state—namely, lack of both a blueprint for assembling a socialist structure, and the material wealth and productive capacity that would allow room for the necessary open-ended experimentation. Marxist literature traditionally had focused on how to overthrow capitalism rather than on what to do after the overthrow. Of course, the USSR did exist as a great world power but, as Mao observed, it was more useful for "learning by negative example" rather than as a model to be duplicated. In the political realm, the official Soviet version of Marxism-Leninism was a frozen variant that rationalized and legitimated the Stalin dictatorship. When transplanted to Third World countries, as it was by multitudinous textbooks, teachers, and ideological mentors, it served to conceal rather than to expose and resolve contradictions. Likewise, the Soviet state lacked the technological and economic dynamism to serve as a model for a new, more communal developmental process. Nor with a few exceptions, most notably Cuba, could the Soviet Union begin to match the investment funds capitalist governments, multilateral institutions, and private banks and investors had available to offer to developing countries. A similar imbalance existed in the cultural field, with no socialist alternatives visible anywhere in the Third World to the music, dress, media productions, and consumer goods gushing out of Hollywood, Madison Avenue, and New York TV shows. "There is more power," French writer Régis Debray has observed, "in rock music, videos, blue jeans, fast food, news networks and TV satellites than in the entire Red Army."[38]

If Third World revolutionary states have received less effective assistance from the Soviet Union than commonly assumed, they have encountered an implacable hostility from the West, whose arsenal of responses include armed intervention, destabilization campaigns, and economic boycotts, as the historical record shows in China, Vietnam, Cuba, Angola, Mozambique, Nicaragua, and other frail young revolutionary states.

The trials and tribulations of these socialist states, however, were by no means entirely foreign in origin. In North Korea, for instance, revolutionary leader Kim Il Sung has created virtually a hereditary monarchy (having designated his son as his successor) in which a cowed population bows down before a monstrous "personality cult." In Vietnam, according to that country's ambassador to the United Nations in 1988, widespread food shortages were caused by "mismanagement, corruption of all kinds, nepotism and incompetence."[39] In Cuba, Fidel Castro visited a new textile factory in 1986 and there denounced a 25 percent absentee rate among workers as "scandalous." He noted that it took twenty-one years to build the factory, and that its roof was still leaky. "We have not taught the people that the first duty of a revolutionary, and the first duty of the citizen, is to work hard and produce, with responsibility and discipline. . . . It is as if we were stuck in a spider web, on taffy or in a swamp. . . . We have to wash the dirty linen."[40]

The result of domestic shortcomings and external assaults is that Third World radicals now find themselves hobbled by institutions and practices virtually the precise opposite of what they originally proclaimed they were setting out to establish. They had anticipated the "withering away" of a state that has only become more controlling and omnipotent. They had planned to replace a class society with a classless society, and yet seemed only to give class a fresh lease on life. They had looked forward to self-reliant socialist economies, independent of the global capitalist order, yet it is those very economies that now are appearing hat in hand at the capitalist doorway and seeking entry, while inviting the capitalist world into their own sanctuaries. How ironic was the observation of Poland's Prime Minister Mieczyslaw Rakowski on the occasion of a November 1988 visit by Britain's ultracapitalist Prime Minister Margaret Thatcher: "I would very much like to be a pupil in her school. I would like to emulate her resoluteness in dealing with unprofitable companies."[41] Equally admiring was Solidarity's Lech Walesa when he said, after thousands cheered Thatcher in Gdansk, that he was "very grateful that fate let me get to know such a fantastic Mrs. Prime Minister."[42]

With both Western capitalism and Eastern socialism having been tried and found wanting, it is scarcely surprising that a growing number in the Third World are turning away from the two great secularist models and toward what is left of their own traditions, which are mainly religious in nature. Manifestations of this turn toward religion are evident on all continents. In Tibet, Buddhist monks have led the protests against Chinese rule; in the Philippines and Haiti, Catholic clergy played prominent roles in the overthrow of local dictators; in Latin America, Catholic "base communities" promote self-help economic projects as well as Bible study with direct social and political implications; while in the Middle East and beyond, even the two superpowers have been cowed by Islamic mujahideen (holy warriors) who forced American Marines to withdraw from Lebanon and Soviet troops to evacuate Afghanistan. "If there is one thing that is happening around the world," states Mexican novelist and academic Carlos Fuentes, "it is the determination of peoples not simply to accept the two versions of inevitable progress—that of Western capitalism or Soviet socialism—but to find ways of combining the power of technology with the energy of their own traditions."[43]

The "traditions" to which Fuentes refers are primarily religious because all the major monotheistic religions preach justice and equality, and both these qualities are conspicuously absent in the contemporary Third World. It follows that those religions are becoming natural focal points of resistance against Third World military dictatorships and exploitative elites. Just as secularist radical political parties make up the opposition in industrialized societies, so in Third World societies the opposition has in recent years tended to crystallize around churches, mosques, and temples.

The two most significant current manifestations of this powerful, if contradictory, religious renaissance are Islamic fundamentalism in the Moslem world, and the theology of liberation in the Catholic world. The intensity and violence of current Islamic fundamentalism stem from the humiliation of centuries of defeats at the hands of the economically and militarily superior West, and more recently at the hands of Israel. Most galling was the 1967 Six-Day War, when Israel defeated three Arab armies and captured extensive territories, including the holy city of Jerusalem. The repeated failures of the Arab states and of the secularist Palestine Liberation Organization (PLO) to reverse this situation have caused many Arabs to turn to Islamic fundamentalism, as faith seemed to provide a more effective basis for mobilizing and motivating people than did nationalism. Fighting in the name of God gives life a deeper sense of mission and worth while escalating the level of struggle to greater intensity because compromise is seen only as a betrayal of God. Whereas the mainstream factions of the PLO have been ready to consider a settlement by which the Holy Land would be divided between Israel and a new Palestinian state, Islamic fundamentalists refuse to consider anything less than the total liberation of the Holy Land. The power of such uncompromising faith was demonstrated in 1985 in southern Lebanon where Shiite Moslems organized a popular resistance that forced the Israelis into a partial withdrawal, a feat that the PLO had failed to achieve anywhere in twenty years of struggle.

Equally impressive was the fusion of Palestinian nationalism and Moslem fundamentalism that was responsible for the *intifada,* or mass uprising, that began in late 1987 in the West Bank and the Gaza area. "The religious awakening of the Arabs in the territories is very threatening," declared Israeli General Amram Mitzna. "If there is something which should bother us in the future, it is a religious awakening, which has begun in the Gaza area and which is growing and liable to intensify."[44] The effectiveness of this fusion of nationalism and fundamentalism has been neatly summarized by Israeli historian Shlomo Avineri: "In 1967 the Israeli Army needed fewer than five days to gain control over the West Bank and Gaza. In 1987 to 1988 the same army—much stronger—cannot restore order when faced with stone-throwing turbulent youths.[45]

Islamic fundamentalism has proven itself not only a powerful force in waging war, but in demonstrating how an alternative Moslem society—one that could exist outside Western influence and control—might be organized. At a time when government services are deteriorating in most Middle Eastern countries, the fundamentalists are opening Islamic schools, hospitals, clinics, small businesses, insurance institutions, and welfare facilities for the destitute.

Even attempting to organize an alternative Islamic society inevitably means challenging the existing order in Arab states throughout the Middle East, a challenge made more serious because of widespread dissatisfaction with regimes considered corrupted by Western money, Western customs, and Western values. Arab students who have studied abroad often return with negative impressions of what they perceive as Western decadence, including broken families, permissive sex, and pervasive materialism. Their reaction is to turn to religion for both physical and psychological sanctuary. Resurgent fundamentalism in the Islamic world reflects an identity crisis—a yearning for cultural and economic as well as political decolonization.

The existing ruling elites, especially in Saudi Arabia and the other Persian Gulf states, feel threatened by the fundamentalist upsurge, and seek to counter it by giving lavish grants to Islamic projects, and by strictly enforcing fundamentalist rules and regulations. These gestures have won little support because the ostentatious life-style of the ruling elites is too visible to be camouflaged, and also because the close ties between the Arab states and the United States, which is viewed as inseparable from Israel, arouse deep suspicion and resentment. Thus, a fundamentalist leader in Gaza states contemptuously that "Arab regimes and Israel are two sides of the same coin."[46] And conversely, a Gulf state ruler asserts equally uncompromisingly that "I would rather deal with ten communists than one Muslim fundamentalist."[47] In line with this appraisal, the American Islamicist James Bill predicts that "over the next forty years, populist Islam is going to be the most important ideological force in the world."[48]

Just as Islamic fundamentalism is a populist religion bubbling up from below in the Moslem world, so the theology of liberation is a populist religion bubbling up from below in the Catholic world. The originators of liberation theology have pictured it as comprising the third stage in the historical evolution of the Catholic Church. During its earliest stage, the church was centered in the Middle East; during the second stage in the Middle Ages its base was in Rome. Today its headquarters, of course, remain in Rome, but 58 percent of all Catholics now live in the Third World, where the church is expanding as it stagnates in most of the West.

As the Catholic Church is being transformed into a Third World church, so liberation theologians hold, church doctrine and structure are bound to change correspondingly. Whereas the church until now has been aristocratic, with decision-making centered in Rome, the future church will be a democratic one in which the laity, overwhelmingly the poor of the Third World, will play a vital and increasingly active role. Instead of an imperial church whose center transmits all its decisions down through the institutional framework, the liberation theologians look toward a federated or polycentric church with scattered centers of decision-making. Whereas the existing church focuses on the religious and the sacred, liberation theologians foresee a church which will be ready to undertake a role as social evangelist in prodding the world toward a long-overdue transformation of global society from the bottom up.

Perhaps the main experimental breakthrough of the liberation theologians has been the emergence of the *comunidades eclesiales de base,* or basic Christian communities (BCCs). These are small groups of people (a dozen to one hundred or more), mostly poor, who combine consciousness-raising, Bible study, worship, mutual help, and political action in defense of their interests. There are an estimated two hundred thousand BCCs in Latin America, of which eighty thousand are in Brazil. Their leaders are priests, nuns, or pastoral agents, the latter coming more and more to the fore. Pastoral agents volunteer their services or receive only enough for subsistence. BCC members read the Bible together, relating it to all aspects of their lives. In the course of their discussions, the poor and the illiterate are meant to discover that they can think for themselves, and discover the truth for themselves rather than having it handed down by others above them. The meetings are held in members' homes or in one-room community centers they have built.

In the course of their meetings, the members seek to improve their quality of life. Traditionally, peasants had tight kinship networks which could be counted on for support when needed by any individual. With so many moving to the cities in recent decades, individuals and nuclear families have been left alone, socially stranded in unknown, alien, and often hostile environments. With the BCCs, the uprooted poor are recreating their support networks. In doing so, and in reading and discussing the Gospel, they see themselves as persons of worth. They develop a new self-confidence, a new conviction that they can free themselves from the state, military, and ecclesiastical bureaucrats and from the landlords and bosses that they had hitherto accepted as virtually divinely ordained masters. The steady spread of this grass-roots movement through Latin America has led an American theologian to conclude that "this vital religious movement is fast becoming the most powerful political force working for change in Latin America—often to the surprise of its own members."[49]

Variants of liberation theology are sprouting in Africa and Asia.

Representatives of these worldwide branches met in Mexico City in December 1986 and adopted measures "to develop a more global concept of liberation theology."[50] In the face of such a grass-roots movement, the papacy, which scarcely welcomes polycentrism in its traditionally centralized institution, has sought to curb the movement by diplomacy and institutional manipulation. Pope John Paul II has traveled incessantly around the globe, reinforcing an image of a powerful church respected by governments and by the global media. On his trips, he has reached out to the poorest of the poor in urban and rural slums, manifesting a deep concern for social justice. His speeches, ethical and moralistic, have fervently denounced exploitation and repression. At the same time, he has proved uncompromising in demanding doctrinal purity and internal discipline.

The pope's church of the future is not a church of the catacombs but a massive, highly centralized, doctrinally unanimous institution. It is no surprise, then, that he has continued to name highly conservative bishops, who will comprise the top ecclesiastical leadership in Latin America for decades to come. He also makes full use of the mass media and all other modern technologies, an example being the $800,000 computer system in Bogotá, Colombia, which keeps a data bank on liberal theologians throughout Latin America, including their speeches, writings, contacts, and travels.

Liberation theology and Islamic fundamentalism each plans to use its holy book to further an egalitarian restructuring of society and to attain the social justice dreamed of by Christ and Mohammed. But the Middle East (and especially Iran) was much less disrupted by Western economic and cultural penetration than Latin America was. Enough of a traditional society remained in place to make plausible the Ayatollah Khomeini's plan for restoring the old order. This is not the case in Latin America, where pre-Columbian Indian societies were smashed beyond recall by the conquistadors, and where Iberian colonial institutions are not held in esteem. So liberation theologians have little choice but to look forward rather than backward. Unlike their Islamic counterparts, they do not say no to both capitalism and socialism. They remain open to new ideas and procedures from both in their groping toward the future.

This explains why there has been a positive interaction between Marxists and liberation theologians that hardly exists between

Marxists and Islamic fundamentalists. Marxists have contributed two key concepts to liberation theology: the need for national economic independence so that local resources can be developed to satisfy local needs; and the need for a restructuring of power in any society in order to mobilize those at the bottom to become active participants rather than remaining passive pawns. Archbishop Helder Cámara of Brazil has concluded that "the Church must do with Marx today what Thomas [Aquinas] did with Aristotle in medieval times."[51]

Conversely, Marxists are learning from Christians, especially about the importance of changing the emphasis in their slogan "democratic centralism" from "centralism" to "democracy." The American theologian Richard Shaull reports that Chilean Marxists, who had held important posts in the regime of Salvador Allende, questioned him closely when they learned that he had information about BCCs in other countries. "They confessed that they had been too elitist. They had thought they had the correct analysis of society and could pass it on to the poor. They realized, however, that they had not been concerned enough to create space *for the poor* in the new society, to empower them, or to set up political structures in which the poor would effectively exercise power. And they admitted that their perspective had changed because of what they had learned about the base communities and through their contact with the work of Catholics in the popular church. They were especially anxious to study this movement and to learn from it."[52]

If a deeper relationship develops between popular Marxist and Catholic movements, the implications for Latin America, not to speak of the entire Third World, are tantalizing. What the end result of those implications might be is impossible to say. Immense confusion and uncertainty prevail today. There is a strong desire to break through to a new noncapitalist vision of how to live, and an equally strong recognition of the failure of the socialist model that had long been accepted as a viable and congenial alternative. Nor does the common legacy from long-gone tributary civilizations provide a way out from this dilemma. The schizophrenic character of that legacy generates more heat than light in the current search for guidelines. The axial religions that were a crucial component of the tributary civilizations were born in periods of social unrest and moral confusion comparable to the present. They offered to the faithful eternal bliss in the afterlife, and brotherhood and communal sharing in this life. The original radicalism of the axial religions was soon tempered as they were co-opted by secular establishments. Buddhist monasteries in China acquired great landholdings and treasures as well as 150,000 slaves who waited on the monks and nuns. Likewise, in the Roman Empire the Christian church became the privileged official religion, and in medieval Europe it became a pillar of the feudal system, acquiring an estimated one-third of all arable land.

Yet the axial religions never shed completely their original egalitarianism, which kept erupting to the surface during periods of social dislocation. Hence, the leadership provided by radical priests during peasant uprisings, such as that of John Ball during the English Peasants' Revolt (1381), and that of Thomas Müntzer in Germany (1525), who denounced "the great" who misused the Law of God to fleece the poor and to protect their ill-gotten gains. "They oppress all people, and shear and shave the poor ploughman and everything that lives—yet if [the ploughman] commits the slightest offense, he must hang."[53] Such incendiary preaching repelled religious leaders as well as secular ones. Martin Luther responded to Müntzer with a ferocious pamphlet entitled *Against the Thievish, Murderous Gangs of the Peasants.* No matter how great the exploitation, declared Luther, rebellion was a heinous sin and a violation of the divinely established order. "I will always side with him, however unjust, who endures rebellion and against him who rebels, however justly."[54]

This exchange illustrates the inherent schizophrenia of institutionalized religion in tributary societies—the deep-rooted and constant tension between revolutionary egalitarian origins and subsequent adaptations and co-options. Like all periods of social disruption and distress, the late twentieth century is witnessing many schizophrenic manifestations. Radical Moslems in the Middle East are proclaiming that to be a true Moslem one must be a revolutionary, while the Saudi dynasty is subsidizing counterrevolution throughout the globe. And radical Catholics in Latin America are proclaiming that to be a true Christian one must be a revolutionary, while Pope John Paul II is admonishing the faithful that "the conception of Jesus as a political figure, a revolutionary, as the subversive from Nazareth, does not tally with the church's catechesis."[55]

It is scarcely surprising that with this historical background and with the ongoing pressures of the global economy, today's Third World is a world trying to break free from what now exists, groping for some way to get from here to there, but with no perception or consensus as to how to traverse the distance to "there" or even as to what "there" should be.

LIFELINES TO A LIVABLE WORLD

All societies are involved to varying degrees in a search for alternative institutions. This is a painful enterprise, undertaken only under the pressure of an even more painful status quo. The pain is becoming intolerable because technology, that distinctive and glorious product of the human mind, has from its very beginnings been abused as much as it has been used.

The use made of technology has been responsible for the matchless success of our human species in the face of seemingly hopeless odds. We first appeared four to five million years ago as puny, relatively scarce, and apparently defenseless creatures confined to the African continent. Today we are the undisputed masters of our planet, and are beginning to extend our sway into the surrounding universe. Our achievements almost defy imagination. Numbering only about 125,000 one million years ago, we now total 5 billion. We have fanned out from our original homeland in Africa and now inhabit the entire globe. We began as food gatherers and scavengers, using sticks and stones as tools in our foraging. Today our tools are computers and space ships. We have harnessed steam, fossil fuels, electricity, nuclear power, and even the sun as energy sources to supplement the human muscles alone available to our early ancestors. We have become the masters not only of the surface of our planet, but now we are learning also the secrets of the outer space of the cosmos and the inner space of the cell.

We humans have been the great overachievers of planet Earth, but we have paid a steep price for our accomplishments, and the reason is to be found in the manner we have used our technological skills. Technological breakthroughs have usually been utilized to benefit the few and to victimize the great majority. The advent of agriculture displaced food gatherers, who made up all of human societies in 8000 B.C. but only 1 percent by 1500 B.C. Likewise, ad-

vanced agriculture and metallurgy, with the resulting state structures and tributary civilizations, divided human communities all
over the world into landed and landless, rulers and ruled, dominant
males and subordinate females. Western industrialization under the
third great societal form, capitalism, and its ensuing overseas expansion, resulted in entire ethnic groups being either exterminated
(Tasmanians and Caribs), decimated and fenced in (Amerindians
and Australian Aborigines) or abducted and enslaved (Africans).
Even within Europe itself, peasants in the East were shackled by the
bonds of serfdom, while in the West they lost their lands to enclosures and were forced to find urban employment or to emigrate as
indentured servants. This process continues to the present day,
though in an accelerated and more intensive way because of the
great power and dynamism of our current High-Technology Capitalism. Peasants on all continents are being uprooted at an accelerating pace. Military technology continues to be employed as an
instrument of national policy, as demonstrated by the havoc of two
world wars and the scores of regional conflicts throughout the globe.

The conclusion to be drawn from this survey of our past is that
we have developed the technological capacity to build a new world,
but have failed to evolve the social capacity for making it a world
worth living in. Even though technology is our own unique creation,
we now face the formidable task of taming it. The genie cannot be
shoved back into the bottle, so the yearning for an uncomplicated
Paleolithic type of existence is destined to remain romantic escap-
ism. Rather, our task is to evolve societal forms capable of turning
our technology into uses beneficial to the majority of humans, and
in a fashion that will nurture rather than devastate the planet we
live on.

Given the nature of the historical record of the last several millenia, it is understandable that many distinguished thinkers have
expressed doubt about the likelihood of that task being fulfilled.
"Man will sooner die than think," observed Bertrand Russell. It is
difficult for an historian, looking at the millions of needless human
deaths in the last hundred years alone, to refute the validity of
Russell's commentary. And yet it does not comprise the whole story.
Along with the negative, analysis of the human past discloses much
that has positive implications. It is necessary to appraise as realistically as possible our legacies from the past in order to understand

the present and prepare for the future. Three such legacies are discernible, emanating from our three great societal forms: Paleolithic kinship bands, tributary civilizations, and capitalist society of modern times.

Our Paleolithic legacy is fundamental because it encompassed by far the larger portion of our history, and because it testifies persuasively that the human species is not genetically programmed to be forever selfish and aggressive. Our Paleolithic ancestors, as noted above, survived the elements and the predators because their cooperative social institutions transformed the basic sex and food drives into bonding rather than divisive drives. The hominid developed into the triumphant *Homo sapiens* because of the ability to share in a complex social context.

The fact that the cooperative communalism of our Paleolithic ancestors contributed fundamentally to their survival is obviously relevant to our own current struggle for survival. It teaches us that what has happened in our historic past was not the inevitable result of human genes but rather of human societies, which, being made by humans, can also in the right circumstances be changed by humans. Far from being haunted by imagined genetic defects that drive humans to act inhumanely, we share instead a proud and beneficent Paleolithic heritage and model. Scientists currently are developing what they term "human needs theory" which holds that "it is simply wrong to claim that human nature is basically individualistic, competitive and aggressive; biologically we are designed to be precisely the opposite. When conflict arises *within* a society, it is almost always because this biologically-based need for bonding among its neighbors is being thwarted by one or another social arrangement."[56]

This is not to suggest that we should imagine ourselves trying to board a time machine to take us back fifty thousand years to the beginnings of our history, without a return ticket. Rather, it is to suggest why the economist Robert Heilbroner in his otherwise somber book *An Inquiry into the Human Prospect* states hopefully that "in our discovery of 'primitive' cultures, living out their timeless histories, we may have found the single most important object lesson for future man. . . . The struggle for individual achievement, especially for material ends, is likely to give way to the acceptance of communally organized and ordained roles."[57]

Our second legacy derives from the tributary civilizations that followed the agricultural revolution, which in turn triggered a great technological leap forward, with its correspondingly great jump in productivity. The resulting surplus was skimmed off by what may be termed the whip of the state—the tax collectors backed by bureaucrats, judiciary and military. Their skimming represented the heavy social price that all civilizations exacted all through history and which included sex discrimination, class differentiation, and institutionalized warfare. But there were positive results also, above all the cultural achievements associated with civilization which we enjoy to the present—achievements in religion, literature, learning, the arts, and crafts.

Great regional civilizations flourished for millennia until over-whelmed by the superior technology and dynamism of Western capitalist states. Nevertheless, the basic characteristics of these civilizations have persisted and are now reasserting themselves. This explains the current awakening of long-dormant groups such as the Basques in Spain, the Armenians in the Caucasus, the Moslem peoples in Soviet Central Asia, and the Celtic peoples in Britain and France. Symptomatic also is the fact that Islam is currently the fastest-growing religion in both the United States and Africa. Tributary civilizations are of more than mere antiquarian interest for today's world. "Encircled nationalisms," writes the Egyptian journalist, Mohammed Heikal, "have fortified themselves for a last ditch stand in the battle for their future, not their past."[58] The battle is under way now, drawing strongly on the tributary legacy, and seems likely to be protracted and pervasive.

Our third legacy is from our capitalist heritage—a heritage of self-generating and accelerating technological breakthroughs cul-minating in the current high technology, which has created for the first time the possibility of non-zero-sum societies. The significance of this achievement can scarcely be exaggerated. All previous soci-eties have been of the zero-sum variety, in that they had available only finite amounts of natural wealth which were claimed and fought over by many contenders, both within nations and between nations. The Roman Empire required 80 percent of its population to work the land in order to marginally feed the full population. Even in the best of times, Roman cities seem to have been about three weeks from starvation conditions. This was a zero-sum situation in that one claimant could get more only by others getting less.

Today this situation has—at least in theory—been reversed in a startling way because the main source of wealth is no longer natural resources but scientific knowledge and technological know-how that has been accumulated incrementally since the advent of civilization. The available pie is potentially no longer finite, and we are potentially no longer trapped in a zero-sum contest. No longer are we forced to pursue a policy of me first and devil take the hindmost. Such a policy may have been profitable in the past, as demonstrated by the fortunes garnered by Alexander the Great in the ancient Near East, by the Spanish conquistadors in the New World, and by British nabobs in India. Today, however, war and conquest are hardly the road to riches, as the Germans and the Japanese discovered during and after World War II. The problem we face now is not the classic postkinship-society one of a finite pie necessitating a struggle for the largest slice. Rather, it is global glut, as evidenced in chronic overproduction, unused capacity, and a multitude of overt and covert trade restrictions designed to keep out the flood of foreign goods seeking markets.

The current global glut may be viewed, in light of the above survey of human history, as a return to the Paleolithic kinship condition of a full larder. We now possess for the first time in ten thousand years the material basis for restoring the full larder, but the circumstances are of our own making rather than of nature's. Kinship relations made possible the free and common utilization of the Paleolithic full larder. The paramount question now is whether we can devise social relationships that facilitate utilization of today's full larder as freely as nature's full larder was in kinship societies.

How that question will be answered is unclear. One reason for the uncertainty is the paradox that just as unprecedented productivity is the great plus of our capitalist heritage, so unprecedented consumption is its unavoidable corollary. The production-consumption spiral began in the eighteenth century when the Industrial Revolution's quantum jump in productivity necessitated an ever-escalating consumerism. The cost of the consumerism is met by the whip of the marketplace, which plays the same skimming role as the earlier whip of the state. But the productivity of capitalist technology is of an entirely different order of magnitude compared to that of the tributary civilizations, and so is correspondingly the efficiency of the whip of the marketplace compared to that of the state. So marked is the difference that questions are being asked with growing frequency and urgency about ecological degradation and social degradation. "Human beings have treated one another badly for as long as we have any historical evidence," declare the authors of *Habits of the Heart*, "but modernity has given us a capacity for destructiveness on a scale incomparably greater than in previous centuries. And social ecology is damaged not only by war, genocide, and political repression. It is also damaged by the destruction of the subtle ties that bind human beings to one another, leaving them frightened and alone. It has been evident for some time that unless we begin to repair the damage to our social ecology, we will destroy ourselves long before natural ecological disaster has time to be realized."[59]

The dilemma posed by the sociologists is compounded by the fact that the whip of the marketplace serves to enforce in a self-perpetuating manner the primary guidelines—"profit or perish" and "produce more or perish." Gandhi has noted the inherent self-destructiveness of this precept: "Earth provides enough to satisfy every man's need but not enough for every man's greed." The relevance of Gandhi's observation was sustained unwittingly by an officer of the Exxon Corporation who asserted that the March 1989 Alaskan oil spill is "the price of civilization."[60] Of course, civilization exacts a price, which we and our ancestors have been paying for millennia. But since the high technology of capitalism and the whip of the marketplace have become so efficient, the price has become correspondingly onerous. So onerous that the dread word "genocide" is being supplanted by even more ominous words—"omnicide" and "geocide." With the future of the human species and of its planetary birthplace at stake, disasters such as the Alaska oil spill, Amazon deforestation, the greenhouse effect, ozone depletion, and toxic dumping on land, sea, and air cannot continue to be dismissed airily as the "price of civilization."

The course of human affairs has reached the point where fundamentals no longer can be ignored, fundamentals such as the purpose of human existence—whether *Homo sapiens* is to end up as *Homo economicus*. The first objective of every society must be to satisfy basic human needs—food, shelter, health, education—so priority must be given to improving economic efficiency until those needs are satisfied. But once they are met, should economic efficiency continue to receive priority regardless of individual, social, and ecological costs? This basic question has not received the consideration it warrants, so that by default a mindless consumerism and materialism is spreading over the planet. As the full implications of this trend become apparent, debate is beginning over those implications and what should be done about them. In forthcoming decades, all peoples throughout the world will have to face up to these issues and participate in the debate. Manifold judgments will evolve slowly and contentiously, in accord with local objective conditions and historical and cultural traditions.

PROSPECTS

What those judgments will be cannot be foreseen precisely, since history is not a crystal-gazing discipline. But history does indicate certain parameters that should be expected and that will have to be reckoned with. One is that no quick fixes will be forthcoming—no instant utopias. Current difficulties in the socialist world have elicited dismay, derision, or impatience. Such reactions reflect an ignorance of comparable birth pangs and growing pains accompanying the historical evolution of capitalism. When that new social system finally took root in England about half a millennium ago, it did so only after centuries of false starts in northern Italy and in the Low Countries. From its birthplace, capitalism spread to the Continent and to the New World, where it assumed distinct regional forms, a trend that continues today as it proliferates throughout the globe.

A comparable slowness and unpredictability have marked the evolution of socialism, whose birth as a system, not a theory, can only be traced back to 1917. The prophets and pioneers of socialism have been as unrealistically enthusiastic as their capitalist counterparts centuries ago, proclaiming confidently the imminent appearance of equitable societies with selfless citizens ("the new Soviet man," "the new Maoist man"). Neither of these male paragons have materialized, nor can any be expected in the foreseeable future. Shortly before his final illness, Lenin himself seemed to recognize the improbability of any quick developments in, much less utopian leaps into, successful socialist systems. He concluded that his emerging new Soviet society was more socialist in appearance than in substance. It was nothing, he said, but the machine "which . . . we took over from Tsarism and slightly anointed with Soviet oil." He added that "the apparatus we call ours is, in fact, still alien to us; it is a bourgeois and Tsarist hotch-potch."[61] Mao was equally aware of the fragility of his handiwork. He expressed this awareness when President Nixon told him during their meeting in Peking in February 1972: "The Chairman's writings moved a nation and have changed the world." Mao replied, "I have not been able to change it. I have only been able to change a few places in the vicinity of Peking."[62]

The historical record not only discourages hopes for quick fixes but also indicates that what we should expect most confidently is the unexpected. Virtually none of the decisive events of recent decades were anticipated, whether the Nazi-Soviet Pact, the A-bombs dropped on Hiroshima and Nagasaki, the post–World War II colonial revolutions, the Soviet-Yugoslav and Soviet-Chinese schisms, Japan's rise to economic leadership, the advent of *glasnost* and *perestroika,* and the current threats of nuclear winter and disruption of our global ecosystem. If we recognize and accept the reality of the unpredictability of future events, then no longer can the lives and liberties of ordinary people be sacrificed in pursuit of certain ends that very likely will prove unattainable.

Instead, today's dogmatists might reflect on Oliver Cromwell's message to the General Assembly of the Church of Scotland on August 3, 1650: "My brethren, I beseech you, in the bowels of Christ, think it possible you may be mistaken." Much more so than in Cromwell's time, it behooves us today to think of the possibility of being mistaken—to guard against any dogma or set of beliefs that we have ensconced above reappraisal. In a world changing infinitely more rapidly than Cromwell's, dogma must give way to working hypotheses constantly reevaluated against unfolding events.

If we must brace ourselves to expect the unexpected, then without in any way being predictive we should take account of an embryonic trend that has at least the possibility of transforming our future in fresh and possibly unimaginable ways—the global participatory impulse. It is an impulse that is astir in all aspects of life—whether in the family, in the classroom, in gender relations, in the workplace, or in politics.

This participatory impulse is not unique to our age. It is discernible throughout history, or at least since the appearance of the state, with its bifurcation of populations into rulers and ruled. Since that bifurcation, the most influential society of each historical period has been the one that reduced the gap between top and bottom—the one that pioneered in raising the level of mass participation. Such quantum jumps constituted the modernity of those pioneering societies, providing them with a qualitatively superior social cohesion and dynamism that enabled them to prevail over all other contemporary societies and to stamp their imprint on their times.

The basic factor behind the "miracle that was Greece" was precisely the leap into modernity effected by the fifth-century Greeks in their city-states. There they attained a level of citizen involvement, ceremonially as well as politically, that had no equal anywhere in the Near East. In the pioneering civilization of Mesopotamia, theology held that "man was . . . created for one purpose only: to serve the gods by supplying them with food, drink, and shelter so they might have full leisure for their divine activities."[63] Altogether different was the attitude of the Greeks, who viewed their gods as being similar in nature to themselves, differing only in superior power, longevity, and beauty. The relationship was essentially one of give-and-take. In return for prayers and sacrifices, the gods were expected to demonstrate their goodwill. Herodotus was very much aware of this difference between Greek culture and that of other Near Eastern peoples, prompting him to refer to their "silliness" in completely subordinating humans to gods.

The Greek leap into modernity helps explain Alexander the Great's triumphal procession from the Aegean to the Punjab, toppling over kingdoms and empires that were overwhelmingly superior in manpower and wealth, but hopelessly inferior because of their social fragility—their fatal breach between top and bottom, between rulers and ruled. The same factor of social cohesion versus social brittleness helps explain why a handful of Spanish conquistadors were able to overturn the Aztec and Inca empires with the indispensable help of thousands of ostensible subjects of these empires. This pattern in which "modernity"—in the form of a more participatory society—prevailed over ossified imperial society was repeated when a few British merchants toppled the Mogul Indian Empire during the seventeenth and eighteenth centuries, again with the aid of native sepoys who did most of the actual fighting. When the Indian mutiny against British rule broke out in 1857, it was put down not only by British troops but also Indian. The correspondent of the London *Times* noted this fact with astonishment, reporting that "men, women and children, with high delight were pouring towards Lucknow to aid the Feringhee [Europeans] to overcome their brethren."[64] The *Times* correspondent was understandably astonished, for the response of his fellow countrymen would have been so completely different if it had been Indians advancing on London. The fact is that the Europeans were able to conquer and rule their great world empires not simply because of their superiority in military technology but equally because of the superior political and social cohesion of their societies.

That cohesiveness was further strengthened when Western legislatures in the late nineteenth and early twentieth centuries expanded the franchise from some males to all males and females, effecting a transition from parliamentary government based on restricted franchise to full representative democracy. Today, however, representative democracy is being challenged in the West (as is the so-called "democratic centralism" in socialist countries). The complaint everywhere is that ostensibly representative institutions in actuality allow only nominal popular participation in government. The demand that has arisen in many confusing and even contradictory forms is for "meaningful" participation—for genuine and substantive input from below in the governmental process. The very concept of "meaningful" participation as envisioned by a Chilean socialist, a Brazilian liberation theologist, a West German Green Party member, or a Middle Eastern Islamic fundamentalist, to take only four examples, may have relatively little in common.

Several factors explain the various current impulses toward participatory as well as representative democracy. One is that modern citizens with more education and access to more information, are more assertive and demanding. Equally important is the fact that representative but nonparticipatory democracy is becoming increasingly nonfunctional. Modern society is far too complex to be directed and controlled solely from above or even from any one power center, whether it be the White House, the Kremlin or 10

Downing Street. No single person or party can effectively run a large modern society any longer, even with all the resources of modern technology. Governments are finding it increasingly difficult to perform even such basic tasks as maintaining law and order, protecting the physical environment, maintaining minimum standards in housing and public health, and providing educational systems suited to national needs.

Still another reason for the current challenge to representative institutions is that they are being exploited for one-way communication through the mass media. Political candidates are merchandised and sold to a market of consumers like soap or toothpaste. "Managers" advised the candidates in the 1988 U.S. presidential elections, for example, to avoid serious discussion of subjects such as foreign policy, the trade imbalance, budget deficits, and arms agreements, and to focus instead on emotionally charged issues such as abortion, capital punishment, prison furloughs, and membership in the American Civil Liberties Union. At the same time, political action committees were short-circuiting the electoral process by using their considerable financial leverage to influence decisively who ran for office and who was elected. American voters have not been unaware of what is going on, as indicated by the steadily declining percentage of eligible voters who have participated in recent presidential elections: 62.8 in 1960; 61.9 in 1964; 60.9 in 1968; 55.2 in 1972; 53.5 in 1976; 52.6 in 1980; 53.1 in 1984; and 50.0 in 1988.

This combination of factors is behind the participatory impulse manifest in varying degrees and forms throughout the globe. In the United States, citizens are turning away from the ballot box because they sense its meaninglessness, yet at the same time they are sharply increasing their participation in day-to-day matters in which they feel they can have some influence. This participatory impulse is now manifest in a broad range of issues: environmental protection, advancement of the rights and role of women, disarmament and peace, food and nutrition, self-education and "self-help." The latter is proliferating rapidly, with no less than five hundred thousand self-help groups functioning in the United States in the mid-1980s and involving over 15 million people. These people have organized in voluntary small-group structures to provide mutual aid in coping with literally the whole range of life crises, from birth to death. Members want to be not merely consumers but also participants in society. They want more control over their lives rather than being dependent on outside professionals. Knowingly or not, they are raising the basic issue of citizen empowerment and are gradually metamorphosing representative democracy into participatory democracy.

The gap between theoretical and de facto self-government is far wider in the socialist Second World than in the capitalist First World. Marxist theory calls for mass involvement and "proletarian democracy," but this theory is negated by the reality of one-party control from above. Just as American voters have questioned the significance of the right to vote when that right is debased by PACs and thirty-second sound bites, so Soviet voters—until Gorbachev's *glasnost*—have been even more skeptical about the purpose of casting ballots in elections with no competing candidates and slates. Gorbachev admitted before a Communist Central Committee meeting on February 19, 1988, that "we have lost and keep losing a lot because of our failure to unshackle grass-roots initiative, endeavor and independence completely. This is the biggest, the hardest, but also the most important task of perestroika."[65]

In an effort to release that "grass-roots initiative," Gorbachev allowed in March 1989 the first contested elections to be held for a new Soviet Parliament. The resulting campaigning whipped up such a political storm that the former *New York Times* Moscow correspondent, David K. Shipler, who was shuttling between Moscow and New York during the election period, concluded that "the quality of political debate has been more intelligent and sophisticated in the world's leading closed society than in the world's leading democracy." In contrast to "the trivial nastiness of the Presidential campaign," Shipler found in Moscow "a creative and exhilarating dialogue on the most fundamental issues concerning the individual and the state."[66] Shipler's conclusion was supported by the election returns which swept to defeat a large number of top party officials, KGB functionaries, and high military commanders. The outpouring of political energy suggests the beginning of a qualitatively new political struggle. Given the opportunity to express their preferences, the Soviet people supported Gorbachev's reforms in that election, yet at the same time they challenged the party and the system that he is seeking to rehabilitate rather than overthrow.

Even more than in the Soviet Union, the participatory impulse has transformed the political landscape in neighboring Poland. Dissidents in that country have been confronted with the overwhelming combined power of the Warsaw and Moscow governments. Any armed resistance or attempt at revolution was foredoomed to failure. Adam Michnik, the underground theoretician and activist, resolved the dilemma of how to satisfy the national yearning for self-determination under such adverse circumstances. In an inspired 1976 essay, he proposed a reversal of the classic revolutionary strategy of seizing state power to effect desired social change. The Polish people, wrote Michnik, should simply ignore the state and proceed to live their own lives, take over their own destiny, and "live as if we were free."[67] The immediate objective should be social rather than political—to build a new alternative society rather than to seize power from the state, and thus to achieve autonomy despite the absence of liberty. Only after de facto self-management had been attained would the time be ripe for negotiation with the authorities.

Michnik, in effect, was calling for the sort of grass-roots initiative that Gorbachev has been seeking with only limited success in the Soviet Union. But in Poland it has been forthcoming because it is in response to a nationalist cause that has popular validity and legitimacy. In recent years under the leadership of the Solidarity trade union, a self-sufficient rival society has taken firm root and is paralleling the Warsaw government. An underground network has proliferated throughout the country, encompassing all aspects of Polish life—from education, health services, and insurance to publishing, the theater, film, radio, and even the production of audio and video tapes. So successful has been the Solidarity strategy based on mass participation that the Polish government found it necessary to sign an agreement in April 1989 that restored the legal status of Solidarity and provided for the free and open election of a two-house parliament in June of that year.

The participatory impulse has been as prominent in the Third World as in the First and Second, though its varying forms reflect the tremendous diversity of local conditions and traditions in that disparate "world." Whatever the nature of the manifestations, positive and negative, the common denominator is spontaneity—an upwelling from below that usually is unexpected in its timing and its magnitude. This is true even of the increase in terrorism in our times.

Terrorism has been customarily viewed as the work of isolated conspiratorial groups supported by hostile governments. An alternative view, however, can be drawn from anthropological research indicating that terrorism more commonly arises relatively spontaneously as a community-based movement. The effectiveness of terrorists in committing their actions and escaping detection is often based directly on a certain degree of support from communities where their cause is seen as righteous and their terrorist acts as virtuous. In this sense, certain communities are described as terrorist-generating, communities such as the Sikhs in India, the Catholics in Northern Ireland, the Shiite Moslems in Lebanon, and the Palestinians throughout the Middle East.

Just as ubiquitous as terrorist acts in the Third World have been the sudden uprisings that have overthrown long-ensconced dictatorships simply through the pressure of overwhelming numbers of almost unarmed aggrieved subjects no longer willing to tolerate oppressive and corrupt rule. Hence, the flight of the shah from Iran, of the Marcoses from the Philippines, and of the Duvaliers from Haiti. Burma's 1988 "people's" uprising, which came close to toppling an entrenched military regime, was described at the time as "one of the few examples of a pure popular revolution that we are seeing anywhere in the world. There are no leaders, there is no organization. . . . a committee of students and monks under 30 years old is maintaining order and performing other functions of government.[68]

Probably the outstanding example of grass-roots activism and open rebellion is under way against the apartheid regime of South Africa. Deep in the black ghettos, "street committees" are being organized that overthrow government-sponsored community councils, drive out the police and their informers, and gradually take over local political power. The black townships are becoming unmanageable; they cannot be governed, they can only be besieged. Street committees throughout the country are coordinating mass-action programs, ranging from consumer boycotts to short general strikes, from revamping school curricula to establishing "people's courts." The grass-roots resistance movement is organized in a hierarchical fashion, with the street committees comprising the base, area committees the next level, then regional committees, and on top an overall coordinating committee. Ultimate power is generally conceded to be wielded by the outlawed African National Congress, whose underground cells provide much of the leadership and organizational strength of the street committee system. "In a situation of ungovernability, the government doesn't have control," states a top African leader, "but neither do the people. While they have broken the shackles of direct government rule, the people haven't yet managed to control and direct the situation. There is a power vacuum. In a situation of people's power, the people are starting to exercise control."[69]

This observation by a present-day African revolutionary brings to mind another revolutionary, Thomas Jefferson, who two centuries earlier on the occasion of the signing of the Declaration of Independence wrote exultantly, "All eyes are opened, or opening, to the rights of man. . . . the mass of mankind has not been born with saddles on their backs, nor a favored few booted and spurred, ready to ride them legitimately, by the grace of God."[70]

Today's revolutionaries are aware of the seditious doctrines espoused by Jefferson and other leaders of the American, French, English, and other western European upheavals. Tanzania's President Julius Nyerere went so far as to lecture the foreign envoys in his capital that Africans were determined to cast off the saddles thrust on their backs, and that no foreign powers had the right to interfere in the liberation process. "The peoples of an individual African country have as much right to change their corrupt government in the last half of the 20th century as, in the past, the British, French and Russian peoples had to overthrow their own rotten regimes. Are African peoples to be denied the same right?"[71]

To anyone in search of historical lifelines, the question naturally arising at this point is whether the participatory impulse, present throughout history, is now reasserting itself on a uniquely massive global scale. Are we witnessing another participatory leap forward surpassing those that in earlier times overwhelmed Charles I, George III, and Louis XVI? The ideology behind those Western upheavals, to which Nyerere alluded, was set forth by John Locke in his *Essay on Civil Government* (1690) in which he stated that if rulers misgoverned their subjects, "by this breach of trust they forfeit the power the people had put into their hands for quite contrary ends, and it devolves to the people, who have a right to resume their original liberty." The important point in this reasoning, however, is that for Locke the "people" comprised educated property owners and not the uneducated and propertyless majority—not the "day labourers and tradesmen, the spinsters and dairymaids" who do not know, who cannot know, and who therefore must be told what to believe and what to do.

The "social contract" that John Locke was demanding represented a leap forward in the seventeenth century. Comparable leaps forward are being demanded today by Solidarity in Poland, by the South African street committees, by the Latin American basic Christian committees, and by the myriad similar organizations active throughout the world. A resurrected John Locke doubtless would be astounded to observe that the term "the people" that he employed in his *Essay* has taken entirely new geographic and social dimensions—that it encompasses inhabitants of all continents rather than of just a few Western countries, and that it includes all "the people" rather than merely a few educated property owners.

Whether the aspirations aroused in the newly activized global masses will be satisfied cannot be foreseen. In the future, as in the past, the unexpected is more likely to materialize. In the process of materialization, those politicized multitudes are probably destined to experience ecstatic highs and heartbreaking lows, as did in their turn the Roundheads and the Cavaliers, the "Sons of Liberty" and the "United Empire Loyalists," the sansculottes and the emigrés, the Bolsheviks and the Mensheviks, the Kuomintang Nationalists and the Maoist Red Guards, and, most recently, the Chinese students and workers camped in Tiananmen Square with their Statue of Liberty and then hunted down throughout China as "counterrevolutionaries" and "ruffians." The only certainty is that just as informed observers in the mid twentieth century could not have foreseen the new technology and new global economic relationships that emerged by the end of their century, so observers today are unlikely to foresee the new politics and new global political relationships that will prevail in the mid twenty-first century.

The tentative status of the human species at this stage of its evolution is reflected in the *Bulletin of the Atomic Scientists*, which publishes on the cover of each monthly issue a "Domesday Clock" whose minute hand registers the level of danger in which humanity is living. For sixteen years, the hand has moved steadily toward midnight. By January 1984, it stood at three minutes before the hour, and it remained there at the brink till February 1988 when the editors moved the hand back to six minutes to midnight. The move reflected the improvement in superpower relations with the elimination of intermediate-range nuclear forces and the Soviet withdrawal from Afghanistan.

The three-minute shift away from midnight may turn out to be an inconsequential upward blip on a long curve ending with Armageddon. On the other hand, the lifelines we have traced from our past allow us to conclude that our age is one of great potentiality as well as of great peril. The lifelines do not force us to either apocalyptic foreboding or to chiliastic fantasies. Rather we reach an open-ended conclusion. Thanks to both our genes and our technology, we are free to be the creators of our destiny rather than the creatures, as some would have it. The choice is ours, and it remains open.

This openness—this spectacle of *Homo sapiens* straddling the globe with one foot planted on top of Mt. Everest and the other in the bottom of Grand Canyon—is what makes this era by far the most exciting and significant in history. Humans today are capable of soaring to new heights as creators or of sinking into oblivion as creatures. The latter course was followed by innumerable species in the past, simply because they were incapable of adapting to their changing environments. Today, humans are themselves responsible for their changing environment, so their problem is not one of adapting to an environment beyond their control but rather one of adapting their own human-made environment to their own human needs.

This would appear to be a relatively easy task for self-styled *Homo sapiens*, yet it remains a task still unengaged because of the constraints of blinders inherited from the past. The roots of this predicament go back several millennia to the time when our Paleolithic ancestors were, in the words of anthropologist Stanley Diamond, dragooned into civilization as conscripts rather than enrolled as volunteers. The dragooning was accepted as necessary because there was no other way at the time by which the burgeoning human family could have been sustained. The resulting worldwide civilizations proved harsh taskmasters, maintained ultimately by the successive whips of the state and of the marketplace. Yet the civilizations did produce results, meeting adequately the needs of their times, which is why they endured for so long. That is also why we have become the great overachievers of our planet. In the process, however, we have become so achievement-oriented that we cannot relent, relax, and enjoy the fruits of past privations and exertions. We are inhibited by blinders that keep us plodding along old routes, unaware of new pathways leading to new highlands and new opportunities. John Maynard Keynes typically pioneered in pinpointing this historic dilemma decades ago when, in the depth of the Great Depression, he labeled unbridled acquisitiveness "a somewhat disgusting morbidity, one of those semi-criminal, semi-pathological propensities which one hands over with a shudder to the specialists in mental disease." Keynes went on to pay tribute to this "morbidity" for being "tremendously useful in promoting the accumulation of capital." Once the "accumulation" has been achieved, concluded Keynes, "we shall once more value ends above means and prefer the good to the useful. We shall honour those who can teach us how to pluck the hour and the day virtuously and well, the delightful people who are capable of taking direct enjoyment in things. . . . it has already begun."[72]

Indeed, it has begun. The question now is whether it will be allowed to continue—whether it will be perceived that self-flagellation and "whips" of any type are now as unnecessary as they are hazardous, and that the time has arrived for those who know how to "pluck the hour" and to "take direct enjoyment of things."

CHAPTER 5

WORLD HISTORY FOR THE TWENTY-FIRST CENTURY

"We enjoy all the achievements of modern civilization that have made our physical existence easier in so many important ways. Yet we do not know exactly what to do with ourselves, where to turn."

VACLAV HAVEL (1994)

Each age writes its own history—not because the earlier history is wrong but because each age faces new problems, asks new questions, and seeks new answers. This precept is self-evident today, when the tempo of change is accelerating exponentially, creating a correspondingly urgent need for new history posing new questions and offering new answers.

Our own generation, for example, was brought up on West-oriented history, and naturally so, in a West-dominated world. The nineteenth and early twentieth centuries were times of Western hegemony in politics, in economics, and in culture. But the two world wars and the ensuing colonial revolutions quickly dismantled the old order, as evidenced by changing names and colors on maps of the world. Slowly and reluctantly, we recognized that, in these altered circumstances, our traditional Western perspective should give way to a global perspective.

Today, convulsed by population explosion, environmental degradation, AIDS, and social-political-ethnic upheavals, the new world taking shape before our eyes is substantially different from that which we inherited following World War II. Consequently, just as the two world wars and the subsequent colonial revolutions necessitated a new globally oriented history, so more innovation in our historical vision is dictated now by ongoing developments.

The issue confronting us is not simply West versus non-West, or a Toynbeean assessment of the meaning of the rise and fall of civilizations. The issue has become more fundamental, involving ourselves as a species—the nature of our human nature, values, achievements, and prospects.

Viewing ourselves as but one link in a long species chain, we are faced with the fact that approximately 40 million different species of plants and animals now exist on planet Earth. Beyond that, somewhere between 5 billion and 40 billion species have existed here at one time or another. Thus, only one in a thousand species is still alive, making the survival record of Earth species a 99.9 percent failure. These statistics reveal the multitude of species that have come and gone on our planet, indicating negligible likelihood of immortality for our own, despite our current global primacy.

A sensitive indicator of human prospects has been the minute hand of the Doomsday Clock appearing in the *Bulletin of the Atomic Scientists* since 1947. Midnight represents the outbreak of nuclear war and the onset of doomsday. The hand has been moved back and forth thirteen times, with the most optimistic setting (at seventeen minutes before midnight) prompted in 1991 by the end of the Cold War and the apparent birth of democracy in the Soviet Union. But in 1995, the *Bulletin* editors moved the minute hand forward once more to fourteen minutes to midnight. The change was made in response to the unremitting turbulence of human affairs, including genocide in Bosnia and Rwanda, proliferating activities of terrorist and paramilitary groups, and growing inequality within and among nations, generating global tensions and conflicts.[1]

The oscillations of the Doomsday minute hand reflect the paradoxical ambivalence of the human condition. *Homo sapiens* currently is scaling the heights of Mount Everest with dazzling achievements in so many areas yet at the same time is haunted by fear of ending at the bottom of Death Valley.

The apprehension concerning human prospects is not limited to atomic scientists. An international team that studied thirty thousand randomly chosen men and women in the United States, Canada, Italy, Germany, France, Taiwan, Lebanon, and New Zealand found that these people are now experiencing major psychological depressions three times more frequently than did their grandparents. Likewise, a study in the United States disclosed that only 1 percent of Americans born before 1905 have suffered a depression by age seventy-five, as compared to 6 percent of those born since 1955.[2]

These statistics suggest a mood of global melancholia, as indicated by the titles of currently circulating books: *End of the American Century, End of the World, End of the Future, End of History*. This is not a unique circumstance, having appeared in past periods of rapid change, which are inherently uncomfortable and stressful. A classic example is the late Roman era, when Saint Cyprian in the third century A.D. was warning of impending and unavoidable catastrophe. He deplored the prevailing corruption, incompetence, and disorder, concluding, "[T]he Day of Judgment is at hand."[3]

Saint Cyprian has many counterparts today, but as historians, we know that he was wholly in error. We know that post-Roman society did not crash into oblivion but rather soared to new heights, leading ultimately to global primacy. We know also that, paradoxically enough, a major factor in this surprising outcome was the barbarian invasions, which had been partly responsible for Saint Cyprian's foreboding. But they also dismantled archaic imperial institutions, clearing the ground for a new civilization destined to dominate the globe in modern times.

Knowing all this, however, is not enough. Our responsibility as historians is not merely to look back knowingly in the clear light of retrospect. We have the responsibility also to look forward—to use our historical knowledge to assess prospects for the future and to prepare for that future. This is especially true now, on the eve of the twenty-first century, when humanity is tottering precariously between Everest and Death Valley.

At this critical juncture, the lifework of the great Italian astronomer Galileo Galilei offers invaluable guidance. In 1608, he constructed an "optic tube," which revealed objects in the night sky thirty times nearer, and a thousand times larger, than ever before. To his "incredible delight," Galileo discovered that the Galaxy "is nothing else but a mass of innumerable stars planted together in clusters."[4] Poets hailed Galileo as a new Columbus, exploring galactic spaces as Columbus had explored oceanic expanses.

Galileo's significance is not so much his construction of the telescope, for he was not a pioneer in that respect. Lens makers in several other European countries were producing increasingly powerful instruments at the time. Galileo's genius was rather in perceiving instantly that his telescope was not merely a toy, as many then viewed it, but an instrument affording unprecedented opportunity for studying and comprehending the heavens as they never had been before. The significance of his vision is evident in Alexander Pope's observation in his "Essay on Man":

> Say what the use, were finer optics given,
> T' inspect a mite, not comprehend the heaven?[5]

An instrument comparable to Galileo's telescope is available to us today in the vast body of historical knowledge accumulated by generations of scholar–scientists. We know more about the ancient civilizations of the Middle East, India, and China than did their own people. Our responsibility is to use that instrument with the insight with which Galileo used his. We need Galileo's insight if we are to avoid the classic misjudgment of Saint Cyprian when he focused on the destructiveness of the barbarians, overlooking their constructive role in clearing the ground for a new post-Roman society.

The parallel between current developments and those witnessed by Saint Cyprian is compelling because, just as the barbarians were then a mighty destructive-constructive force, so today a comparable force is playing a comparable role—namely, our high technology. This comprises the bundle of breakthroughs stimulated largely by World War II—breakthroughs such as labor-replacing machinery, genetic engineering, space science and exploration, and revolutionary advances in accumulating and distributing information.

Just as successive waves of Vandals, Visgoths, and Huns cleared the way for modern civilization, so the recurring waves of ongoing high technology are undermining existing civilization and clearing the way for a new twenty-first–century civilization. This proposition is not as fanciful as it may first appear if it is recognized that the disruption now inflicted on all societies by high technology is far greater than that inflicted by the barbarians on the Roman Empire.

Example: Ethnic Disruption. The barbarian invasions changed the ethnic map of Europe, with Germans settling down in the north, Slavs in the east, and Magyars in the center. Comparable ethnic rearranging is occurring today, though on a vastly greater worldwide scale. It is triggered by modern steamship and airline facilities as well as by multinational corporations whose global business operations require a global labor force. Hence, the great migration currents now under way: to the United States from Latin America and Asia, and to Northern Europe from Southern Europe, North Africa, South Asia, and the West Indies. The scale of these migrations is such that the United States, which once was unique as a "melting pot" of Europeans, now is again becoming "unique" as a "world nation," with immigrants, legal and illegal, pouring in from all sides. So far-reaching is the ethnic shift that some foresee a change in the complexion of Americans. "We are probably going to have a browning of America over time. It will take six or seven generations," declares an expert on the subject, "but ultimately, I believe a majority of the population will be nonwhite."[6]

Example: Political Disruption. The barbarians dismantled the Roman Empire, but high technology is largely responsible for the collapse of the Soviet economy and for the subsequent disappearance of the U.S.S.R. from the world map. Other states are also being buffeted by high technology, specifically by multinational corporations that operate without regard to political frontiers, thanks to new technological facilities. These include containerized shipping, which lowers transportation costs; combined computer and satellite communication for global coordination of production; and computerized cash management systems for tapping world money markets. So rapidly have multinational corporations grown that they are overshadowing nation–states and becoming increasingly independent of them. Indicative of this trend is the Citicorp chairman's prediction that a few large financial institutions soon will dominate the world economy and will have their headquarters on space platforms in order to "avoid government regulations."[7]

Example: Ecological Disruption. Parallel to the disruption of societies is the ongoing disruption of planet Earth. This occurs partly because of the development of human technology from Paleolithic sticks and stones to computers and spaceships. Equally important is the exponential growth in population and economic activity. In the twentieth century alone, world population increased threefold, gross world product twentyfold, and fossil fuel use tenfold. These increases have generated corresponding escalation of stresses on the planetary ecosystem, including the greenhouse effect; pollution of land, sea, and air; deforestation; and desertification, which, according to a U.N. study, is threatening 35 percent of the Earth's surface, with the Sahara Desert expanding six to twelve miles per year. Because of this formidable combination of planetary stresses, a private environmental body, the World Watch Institute, concluded in 1989 that our generation is "the first to be faced with decisions that will determine whether the earth our children inherit will be habitable.... By the end of the next decade, the die will pretty well be cast. As the world enters the twenty-first century, the community of nations either will have rallied and turned back the threatening trends, or environmental deterioration and social disintegration will be feeding on each other."[8]

Example: Economic Disruption. Equally disruptive is the economic impact of high technology, with labor-*replacing* innovations succeeding the earlier labor-*saving* instruments. The implication of this development, states philosopher Herbert Marcuse, is that it "threatens to render possible the reversal of the relation between free time and working time, [with] working time becoming marginal and free time becoming full time. The result would be a mode of existence incompatible with traditional culture."[9]

Confronted with such global disruption, the reported global melancholia may appear understandable. But not necessarily justifiable, would add the historian who aims his or her optic at the heavens rather than at mites. That historian would note that high technology, like the barbarian invasions, has a pervasive constructive as well as negative impact.

One negative impact deeply imprinted on the human mind has been the global arsenal of fifty thousand nuclear weapons accumulated during the Cold War. In November 1983, a consortium of scientists from several countries warned that if only a small fraction of those weapons were detonated, it would precipitate a "nuclear winter." Firestorms and massive amounts of smoke, oily soot, and dust would blot out the sun and plunge the Earth into a freezing darkness for from three months to a year or more. "Global environmental changes sufficient to cause the extinction of a major fraction of the plant and animal species on the earth are likely. In that event, the possibility of the extinction of *Homo sapiens* cannot be excluded."[10]

Such "extinction" was, and still is, a dreaded possibility. Yet the fact remains that Carl Sagan, one of the space scientists who warned of the danger of nuclear winter, sounded a hopeful note since that warning. In a more recent study, *Pale Blue Dot: A Vision of the Human Future in Space* (Random House, 1994), Sagan noted that the nuclear missiles hanging over our heads like so many Damoclean swords may also serve as a protective shield against an even greater danger. This is the multitude of asteroids circling the Earth—more than the number of stars visible to the naked eye. Periodically, they crash into Earth, and if the crashing object is seventy meters in diameter, it releases energy equal to the largest nuclear weapons explosion ever detonated. This considerable peril to human life, noted Sagan, can be countered by tracking the orbits of the asteroids, determining which are on a collision course with Earth, and deflecting them with blasts from nuclear missiles—all of these procedures being well within the range of current technology and paralleling its destructiveness.

Turning to more mundane concerns, high technology has enabled humans for the first time to break free from the restraints of what economists call zero-sum civilizations. Those civilizations had available only finite amounts of natural wealth, which were claimed and fought over by many contenders, both within nations (class wars) and among nations (state wars). This was a zero-sum situation, in that one claimant could get more only by others' getting less. Today, the situation is precisely the opposite because the main source of wealth is not finite natural resources but ever-expanding scientific knowledge and technological know-how.

Consequently, humans are no longer trapped in a zero-sum contest for limited resources. Rather, they face the novel problem of a global glut, with world industries and farms producing more than world consumers can absorb or can afford to absorb. The reality of the glut is apparent in the multitude of overt and covert trade barriers, despite the official rhetoric about global free trade. This is a welcome respite from the traditional economic scarcity, exemplified by the Roman Empire, whose cities always seem to have been about three weeks from starvation.

The superproductivity of high technology has the negative effect of runaway consumerism, exacerbated by unrelenting advertising campaigns. But the consumerism, in turn, has generated a proliferating counterphenomenon known variously as Voluntary Simplicity, Downshifting, and Simple Living. This reflects a shift in values and priorities, by people satiated by consumerism and yearning for more free time and less stress. Harvard economist Juliet B. Schor, author of *The Overworked American*, views these people, who are voluntarily reducing their workload and paycheck, as individuals who are saying, " 'I no longer want to sacrifice my time, my sanity, my stress level to make money.' . . . What they are doing is experiencing a change in values, or new priorities on what's important."[11]

High technology's productivity generates ecological problems as well as personal and social ones. The progression from Paleolithic sticks and stones to nuclear plants and spaceships has left its stamp on the planetary ecosystem, a stamp greatly magnified by the simultaneous sharp growth in population and economic activity.

Once more, excess has stimulated corrective antidotes, so that spy satellites formerly peering down at military targets are now monitoring natural phenomena such as clouds, glaciers, sea ice, deserts, and tropical rain forests. Similarly transformed is the world energy economy, long dominated by large oil refineries, internal combustion engines, and steam-driven power plants. With consumers demanding a cleaner environment and cheaper energy, new technologies are harnessing the world's most abundant energy resources—the sun and wind. Silicon cells, turning sunlight directly into electricity, are being installed in remote areas lacking power lines. Likewise, new fiberglass wind turbines with gearless variable-speed transmissions and advanced electronic controls are now generating electricity more cheaply in many regions than do coal plants. In the United States, all of the country's electricity could be supplied by turbines in three states (North Dakota, South Dakota, and Texas), and likewise, Inner Mongolia could meet all of China's electricity needs. The latter country also has available another plentiful new energy source in the form of biogas produced from human and animal wastes. To build their own power plant, peasants need only a shovel, a pig, and a latrine. As an added bonus, this poor people's petroleum has wiped out North China's endemic intestinal worms and snail fever, diseases transmitted by untreated sewage tossed on open fields.

All these ongoing developments make clear how high technology is now scouring and transforming the globe as thoroughly as the barbarians did the Roman Empire. New potentialities are emerging, together with new visions of the twenty-first century. "I see all sorts of wonderful, creative things happening around the world," declares Willis Harman, Stanford engineering professor and systems analyst, "with people creating alternative economies, new kinds of entrepreneurship, new kinds of communities. . . . Ours is one of the great times of human existence when we're making an evolutionary leap that people will span in a single lifetime."[12]

Cal-Tech geochemist Harrison Brown is similarly exuberant about human prospects. "I am just as convinced as can be that man today has much more power than he realizes. I am convinced that man has it within his power to create a world in which people the world over can lead free and abundant and even creative lives. . . . I am convinced that we can create a world which will pale the Golden Age of Pericles into nothingness."[13]

It is tempting at this point to make the comforting assumption that because revolutionary medieval technology created modern civilization, so today's more revolutionary high technology will create a new twenty-first–century civilization. The analogy is intriguing and is sustained by the above jubilant appraisals. But with humans continually introducing unpredictable variables, history does not repeat itself so mechanically.

It is not only human activities that are unpredictable; equally so is the inherent dynamics of technology. The distinguished biologist David Suzuki states flatly, "[T]here is no such thing as cost-free technology. The more powerful the technology, the greater the probable cost, and the more likely it is to be unpredictable."[14] A classic example of this proposition is to be found in one of the earliest human technological innovations, the Agricultural Revolution. Agriculture proved an immediate and priceless boon, providing the extra food needed by extra mouths. But in solving the problem of food scarcity, the Agricultural Revolution generated three new, unforeseen social problems—population explosion, ecological deterioration, and social inequity.

The significance of these three problems is that they persist to the present day and have festered during the intervening ten thousand years to the point of endangering us and even our planet. This fact raises the fundamental question of why, through the millennia, we have been mired in social stagnation amid technological storms of our own making. The answer is to be found in our cultures. All cultures of all peoples consist of control mechanisms designed to regulate the behavior of society members. The mechanisms evolved gradually during the historical evolution of the societies and therefore represented their survival wisdom. The values comprising the various cultures were calculated to enhance community cohesion and survival. Consequently, the values commonly incorporated in cultures favored maximum fertility for species perpetuation, maximum productivity for economic sustenance, and maximum military strength for physical survival.

Through the millennia, cultures became the essential underpinnings of their respective societies. Only through their cultures did individuals know what they could do and how to do it. Therefore, any threat to cultural values became as serious as any threat to other basic necessities such as food and water. Hence, the extreme reluctance to tolerate any substantive alteration or modification of traditional values. Hence, also the historic persistence of cultural rigidity or inability to cope with societal afflictions such as population explosion, environmental degradation, and social inequity generated by the Agricultural Revolution ten thousand years ago.

This elemental role of cultures in human history has been pinpointed vividly by the poet William Blake, with his inspired phrase defining cultures as "mind-forged manacles." Being forged by the mind, those manacles have proved infinitely more durable than those forged from any metal. They have endured from earliest times because they have endorsed the familiar and comfortable and repudiated the unfamiliar and uncomfortable. President John F. Kennedy perceived clearly this implicit constraint exerted by the manacles: "Too often we hold fast to the clichés of our forbearers. We subject all facts to a prefabricated set of interpretations. We enjoy the comfort of opinion without the discomfort of thought." The English philosopher Bertrand Russell concurred emphatically: "Man will sooner die than think."

The validity of Russell's judgment is manifest in the hundreds of millions who have died needlessly in past centuries. It is manifest more recently in the birth pangs of modern civilization when it was emerging out of its medieval context. The church traditionally had denounced interest charges as constituting usury, a mortal sin and "a vice most odious and detestable in the sight of God." But this stance became increasingly burdensome and unpopular with overseas expansion offering opportunities for lucrative commercial ventures and profiteering. Church members soon were pleading for the acceptance of "moderate and acceptable usury." Finally, the tension between the manacles of the past and the allure of the market prompted a French banker to assert resentfully, "He who takes usury goes to hell; he who does not goes to the poorhouse."

Likewise, executives today, constrained by "bottom-line" pressures, are resisting demands for reduced work hours and for increased wages commensurate with productivity increases. Hence, the current anomalies of growing inequity and privation in an era of glut and of overwork in an era of computers and robots. Such anomalies were glossed over in the past, but this is no longer feasible with runaway population increase and technological proliferation and with the attendant ecological consequences. Oceanographer Jacques Cousteau has warned, "Mankind has probably done more damage to the Earth in the 20th century than in all previous human history."[15]

Cousteau pinpoints responsibility for the damage in an article entitled "Consumer Society Is the Enemy." He describes a one-day walk in Paris, from 7:00 A.M. to 7:00 P.M. During his walk, he had a counter, which he clicked "every time I was solicited by any kind of advertising for something I didn't need. I clicked it 183 times in all by the end of the day." Cousteau concludes from this experience: "It is the job of society, not of the individual person, to control this destructive consumerism. I am not for some kind of ecological statism. No. But when you are driving in the street and see a red light, you stop. You don't think the red light is an attempt to curb your freedom. On the contrary, you know it's there to protect you. Why not have the same thing in economics? . . . Responsibility lies with the institutions of society, not in the virtues of the individual."[16]

Cousteau's conclusion raises basic questions for both societies and individuals. His hypothesis was a central issue during the decades of the Cold War. It was not settled with the disappearance of the U.S.S.R. because factories all over the world continue to churn out consumer goods, and advertising continues to stimulate popular demand for those goods. The demand is becoming worldwide, as is now evident in China. When Mao Tse-tung began his rule in 1949, the popular clamor was for the "big four" (bicycle, radio, watch, and sewing machine). Since then, consumer expectations have escalated to the "big eight," adding items such as color TV, refrigerator, and motorcycle.

The list continues to lengthen, an outstanding recent addition being the automobile, which is becoming a status symbol among the billions of "have-nots" in the Third World. Between 1990 and 2000, the number of automobiles will increase in Indonesia from 272,524 to an estimated 675,000; in India from 354,393 to 1,100,000; and in China from 420,670 to 2,210,000.[17]

Environmentalists are concerned about the impact of millions of additional refrigerators on the ozone layer and of millions of additional automobiles on the global atmosphere. Norway's former prime minister Gro Harlem Brundtland notes, however, that Western Europeans who initiated the Industrial Revolution and the ensuing atmospheric pollution cannot now condemn "have-nots" to the status of "never-will-haves."[18]

These circumstances raise profound issues for individuals and for societies, today and in the foreseeable future. So we find ourselves at a point where we can no longer avoid facing up to fundamentals. What is the meaning of life? What is the purpose of human existence? Francis Bacon faced up to this centuries ago when he urged that the newly emerging discipline of science be employed for the "benefit and use of life" and not for "inferior things" such as "profit, or fame, or power."[19] Bacon posed the issue squarely: must *Homo sapiens* end up as *Homo economicus,* dedicated to achieving a bloated stomach and a bloated bank account?

The first objective of every society must be to satisfy basic human needs—food, shelter, health, education—so priority must be given to improving economic efficiency until those needs are satisfied. But once they are met, should economic productivity continue to receive priority regardless of individual, social, and ecological costs? This basic question has not received the consideration it warrants, so that by default, a mindless consumerism and materialism has spread over the planet.

Such equivocation cannot be sustained indefinitely. So *Homo sapiens* now is engaged willy-nilly in a search for an alternative to *Homo economicus,* or more precisely, for an ethical compass to direct our rampant technology. Participation in this search is the manifest role of twenty-first–century world history. Thus, we are now presented with an opportunity to play a role comparable to that of Galileo.

The opportunity is outstanding, as is the responsibility. To grasp it requires Galileo's courage in confronting detractors and enduring imprisonment for unorthodox views and also his vision in focusing his optic upward to the stars rather than downward to the mites.

NOTES

LIFELINES FROM MY PAST

1. P. Kennedy, *The Rise and Fall of the Great Powers: Economic Change and Military Conflict from 1500 to 2000* (Random House, 1988).
2. Cited by R. Maheu in *UNESCO Courier*, February 1966, p. 30.
3. Cited in E. Fromm, *The Anatomy of Human Destructiveness* (Holt, Rinehart & Winston, 1973), p. VII.
4. Most recent are this author's *A Global History: From Prehistory to the Present*, 4th ed. (Prentice-Hall, 1988); W. H. McNeill, *A History of the Human Community: Prehistory to the Present*, 2nd ed. (Prentice-Hall, 1987); and H. Thomas, *A History of the World* (Harper & Row), 1979). Noteworthy in this connection is the admirable *Ariadne's Thread: The Search for New Modes of Thinking* by Mary E. Clark (St. Martin's, 1989) an audacious and informative effort by a biologist to explore the "new modes of thinking" exhorted decades ago by Albert Einstein.

1. KINSHIP SOCIETIES

1. C. Stringer, "The Dates of Eden," *Nature*, February 18, 1988, pp. 565–66, 614–16; and R. Lewin, "Modern Human Origins Under Close Scrutiny," *Science* 239 (March 11, 1988), pp. 1240–41.
2. E. R. Service, *The Hunters* (Prentice-Hall, 1966), p. 14.
3. P. Freuchen, *Book of the Eskimos* (World Publishing Co., 1961), p. 154.
4. B. Spencer and F. J. Gillen, *The Arunta* vol. I (Macmillan, 1927), p. 37, cited in Service, *op. cit.*, p. 17.
5. R. Lewin, "A Revolution of Ideas in Agricultural Origins," *Science* 240 (May 20, 1988), pp. 984–86.
6. Service, op. cit., p. 32.
7. M. D. Sahlins, "The Origins of Society," *Scientific American* 203, no. 3 (1960), pp. 80, 86.
8. V. Stefansson, "Lessons in Living from the Stone Age," in H. Shapley, ed., *A Treasury of Science* (Harper, 1943), pp. 510–11.
9. Thomas Hobbes, *Leviathan* (London, 1651).
10. Sahlins, *Stone Age Economics* (Aldine-Atherton, 1972), p. 34.
11. R. B. Lee, *The !Kung San* (Cambridge University Press, 1979), p. 246.

12. M. Shostak, *Nisa: The Life and Words of a !Kung Woman* (Harvard University Press, 1981), p. 16.

13. S. Diamond, *The Search for the Primitive* (Dutton, 1974), p. 138.

14. Sahlins, *Stone Age Economics,* p. 33.

15. Freuchen, op. cit., pp. 145–54.

16. *Los Angeles Times,* December 22, 1986.

17. Freuchen, op. cit., p. 178.

18. A. T. Rambo, *Primitive Polluters: Semang Impact on the Malaysian Tropical Rain Forest Ecosystem* (Anthropological Papers, Museum of Anthropology, University of Michigan, no. 76, 1985), pp. 36–42.

19. Peter Nabokov, ed., *Native American Testimony, An Anthology of Indian and White Relations: First Encounter to Dispossession* (Harper & Row, 1979).

20. Cited by E. Leacock, "Women in Egalitarian Societies," in R. Bridenthal and C. Koong, *Becoming Visible: Women in European History* (Houghton Mifflin, 1977), pp. 21–22.

21. Leacock, "Class, Commodity and the Status of Women," in S. Diamond, ed., *Toward a Marxist Anthropology* (Mouton Publishers, 1979), p. 193.

22. Cited in H. Reynolds, *The Other Side of the Frontier* (Penguin, 1982), p. 151.

23. *Los Angeles Times,* December 22, 1986.

24. Lee, op. cit., pp. 460–61.

25. R. G. Thwaites, ed., *The Jesuit Relations and Allied Documents,* vol. III (Cleveland, Burrows, 1897), pp. 84–85, cited in P. Farb, *Humankind* (Houghton Mifflin, 1978), p. 96.

26. Sahlins, *Tribesmen* (Prentice-Hall, 1968), p. 79. Emphasis in original.

27. G. A. Zegwaard, "Headhunting Practices of the Asmat of Netherlands New Guinea," *American Anthropologist* 61 (December 1959), p. 1040.

28. Cited by W. Goldschmidt, "Personal Motivation and Institutionalized Conflict," in M. L. Foster and R. A. Rubinstein, eds., *Peace and War: Cross-Cultural Perspectives* (Transaction Books, 1986), p. 6.

29. Ibid.

30. E. R. Service, "War and Our Contemporary Ancestors" in M. Fried et al., *War: The Anthropology of Armed Conflict and Aggression* (Natural History Press, 1968), p. 160.

31. A. Bandura, *Aggression* (Prentice-Hall, 1973), pp. 113, 322.

2. TRIBUTARY SOCIETIES

1. A. S. McNeilly, "Effects of Lactation on Fertility," *Medical Bulletin XXXV* (1979), pp. 151–54.

2. A. I. Richards, *Land, Labour, and Diet in Northern Rhodesia* (Oxford University Press, 1939), pp. 162–64.

3. E. R. Wolf, *Peasants* (Prentice-Hall, 1966), pp. 5–10.

4. J. I. Lockhart, trans., *The Memoirs of the Conquistador Bernal Díaz de Castillo,* Vol. I (J. Hatchard, 1844), pp. 220–23, 228–41.

5. Leo Africanus, *A History and Description of Africa* (Hakluyt Society, 1896), p. 825.

6. T. Hodgkin, "Islam in West Africa," *Africa South* II (April–June 1958), p. 98.

7. Cited in E. Galeano, *Memory of Fire: Genesis* (pt. 1 of a trilogy) (Pantheon, 1985), p. 158.

8. J. Needham, *Science and Civilisation in China,* vol. I, (Cambridge University Press, 1954), pp. 209–10.
9. Cited in M. Selden, *The Yenan Way in Revolutionary China* (Harvard University Press, 1971), pp. 191–92.
10. *Los Angeles Times,* April 19, 1986.
11. D. C. Lau, trans., *Mencius* (Penguin, 1970), p. 92.
12. Lynn White, Jr., "The Life of the Silent Majority," in R. S. Hoyt, ed., *Life and Thought in the Early Middle Ages* (University of Minnesota Press, 1967), pp. 85, 86.
13. Plato, *Critias,* III.
14. Cited in T. Dale and V. G. Carter, *Topsoil and Civilization* (University of Oklahoma Press, 1955), p. 6.
15. W. C. Lowdermilk, *Palestine: Land of Promise* (Victor Gollancz, 1944), p. 25.
16. Cited in S. Pomeroy, *Women in Classical Antiquity* (Schocken, 1976), p. 8.
17. Cited in E. Leacock's introduction to F. Engels, *The Origin of the Family, Private Property, and the State* (International Publishers, 1972), p. 38.
18. Cited in A. F. Wright, *Buddhism in Chinese History* (Stanford University Press, 1959), pp. 19–20.
19. M. Loewe, *Everyday Life in Early Imperial China* (Putnam, 1968), pp. 138–41.
20. K. C. Chang, ed., *Food in Chinese Culture* (Yale University Press, 1977), p. 15.
21. L. H. Morgan, *Houses and House-life of the American Aborigines* (New York, 1881), p. 45.
22. W. A. Haviland, "Stature at Tikal, Guatemala," *American Antiquity,* July 1967, pp. 316–25.
23. *New York Times,* December 11, 1984.
24. Cited in E. Sagan, *At the Dawn of Tyranny* (Knopf, 1985), p. 281.
25. Cited in J. Blum, *Lord and Peasant in Russia* (Princeton University Press, 1961), p. 556.
26. Cited in W. B. Walsh, ed., *Readings in Russian History* (Syracuse University Press, 1950), pp. 205–6.
27. Cited in Blum, op. cit., p. 566.
28. Adapted from J. Hawkes and L. Wooley, *Prehistory and the Beginnings of Civilization,* UNESCO History of Mankind series, vol. I (Harper & Row, 1963), p. 467; and V. Gordon Childe, *Man Makes Himself* (New American Library, 1951), p. 149.
29. "Work and Leisure in Pre-Industrial Society," *Past + Present,* no. 29 (December 1964); "Work and Leisure in Industrial Society," *Past + Present,* no. 31 (July 1965).
30. N. Cohn, *The Pursuit of the Millennium* (Essential Books, 1957), p. 214.
31. *The Travels of Marco Polo* (Oliver & Boyd, 1844), p. 191–93.
32. Lockhart, trans., op. cit., pp. 228–41.
33. M. Sahlins, *Stone Age Economics* (Aldine-Atherton, 1972), p. 37. Emphasis in original.
34. Cited in N. Bailkey, "Early Mesopotamian Constitutional Development," *American Historical Review* LXXII (July 1967), p. 1225.
35. Cited in G. Dyer, *War* (Crown, 1985), p. 4.
36. J. A. Giles, trans., *Matthew Paris' English History,* vol. I (London, 1852), pp. 312–13.

3. CAPITALIST SOCIETIES

1. J. A. Schumpeter, *Capitalism, Socialism, and Democracy,* 3rd ed. (Harper & Row, 1950), p. 83.
2. J. M. Keynes, *Essays in Persuasion* (Harcourt, 1932), pp. 360–61.
3. Schumpeter, op. cit., p. 82.
4. Cited in J. Needham, *Science and Civilization of China,* vol. 4, pt. III: *Civil Engineering and Nautics* (Cambridge University Press, 1971), p. 584.
5. *Old South Pamphlets* 2, no. 33 (Directors of the Old South Work, 1897).
6. T. K. Rabb, "The Expansion and the Spirit of Capitalism," *Historical Journal* XVII (1974), p. 676.
7. W. Eton, *A Survey of the Turkish Empire,* 4th ed. (London, 1809), pp. 190–93.
8. T. Sprat, *The History of the Royal Society of London* (London, 1734), p. 72.
9. Cited in F. L. Baumer, ed., *Main Currents of Western Thought* (Knopf, 1954), p. 251.
10. E. Baines, *History of the Cotton Manufactures in Great Britain* (London, 1835), pp. 5–6.
11. H. S. Dinerstein, "The Sovietization of Uzbekistan," *Russian Thought and Politics* (Harvard Slavic Studies IV, 1957), p. 503.
12. H. W. Temperley, "Causes of the War of Jenkins' Ear," *Royal Historical Society Transactions,* 3rd series, III (1909), pp. 197–206; F. Whyte, *China and Foreign Powers* (Oxford University Press, 1927), p. 38.
13. F. Wakeman, Jr., *The Great Enterprise: The Manchu Reconstruction of Imperial Order in Seventeenth-Century China,* vol. I, (University of California Press, 1985), pp. 2–4.
14. *The Log of Christopher Columbus' First Voyage,* cited in E. Galeano, *Open Veins of Latin America* (Monthly Review Press, 1973), p. 24.
15. Cited in E. Reynolds, *Trade and Economic Change on the Gold Coast, 1807–1874* (Longman, 1974), p. 42.
16. Ibid., p. 43.
17. Cited in M. Beaud, *A History of Capitalism 1500–1980* (Monthly Review Press, 1983), p. 32.
18. Cited in *The Encyclopaedia of Islam,* vol. III, new ed. (Brill, 1971), p. 1187.
19. W. Lippmann, *Preface to Morals* (Macmillan, 1929), p. 235.
20. W. H. McNeill, *The Pursuit of Power, Technology, Armed Force, and Society Since A.D. 1000* (University of Chicago Press, 1982), chap. 8.
21. Cited in O. C. Cox, *Capitalism as a System* (Monthly Review Press, 1964), p. 173.
22. Cited in K. O. Dike, *Trade and Politics in the Niger Delta 1830–1885* (Clarendon Press, 1956), p. 211.
23. Cited in L. Huberman, *We, The People,* rev. ed. (Harper, 1947), p. 263.
24. W. T. Stead, *The Last Will and Testament of Cecil John Rhodes* (London, 1902), p. 190.
25. Cited in S. Marcus, *Engels, Manchester, and the Working Class* (Random House, 1977), p. 66.
26. Cited in F. Clairmonte, *Economic Liberalism and Underdevelopment* (Asia Publishing House, 1960), p. 86.
27. Hwuy-ung, *A Chinaman's Opinion of Us and His Own Country,* trans. J. A. Makepeace (Chatto & Windus, 1927).

28. Cited in L. S. S. O'Malley, *Modern India and the West* (Oxford University Press, 1941), p. 766.
29. S. Avineri, ed., *Karl Marx on Colonialism and Modernization* (Anchor Books, 1969), pp. 89, 94.
30. *Democratic Values*, Fabian Tract no. 282 (London, 1950), cited in G. Myrdal, *Rich Lands and Poor* (Harper, 1957), p. 46.
31. *New York Times*, April 2, 1949.
32. Cited in S. and B. Webb, *Soviet Communism: A New Civilization*, vol. II (Victor Gollancz, 1937), p. 605.
33. M. Lewin, *The Gorbachev Phenomenon: A Historical Interpretation* (University of California Press, 1988), pp. 15–18.
34. *Toward Freedom: The Autobiography of Jawaharlal Nehru* (Day, 1941), pp. 229–30.
35. A. J. Toynbee, *Survey of International Affairs, 1931* (Oxford University Press, 1932), p. 1.
36. C. P. Snow, "Government, Science, and Public Policy," (Committee on Science and Astronautics, U.S. House of Representatives, 89th cong., 2nd sess., 1966), p. 7.
37. R. N. Bellah et al., *Habits of the Heart* (Harper & Row, 1985), p. 284.
38. T. Von Laue, *The World Revolution of Westernization* (Oxford University Press, 1987), p. 240.
39. P. F. Drucker, "The Changed World Economy," *Foreign Affairs*, Spring 1986, pp. 770–74.
40. Cited in H. Stephenson, *The Coming Clash* (Weidenfeld & Nicolson, 1972), p. 12.
41. I. Wallerstein, "The World-System: Myths and Historical Shifts," in E. W. Goldolf et al., eds., *The Global Economy: Divergent Perspectives on Economic Change* (Westview Press, 1986), pp. 20–21.
42. J. Nehru, *Toward Freedom* (Beacon Press, 1958), p. 264.
43. Cited in E. G. Vallianatos, *Fear in the Countryside* (Ballinger Publishing Co., 1976), p. 100.
44. *Washington Spectator*, May 1, 1987, p. 4.
45. *Los Angeles Times*, April 17, 1986; April 29, 1986.
46. Drucker, op. cit., p. 775.
47. *New York Times*, March 26, 1986; August 4, 1987.
48. *New York Times* and London *Times*, April 3, 1970.
49. J. K. Galbraith and S. Menshikov, *Capitalism, Communism, and Coexistence* (Houghton Mifflin, 1988), p. 32; L. S. Stavrianos, *Global Rift* (Morrow, 1981), pp. 85–90.
50. *Los Angeles Times*, June 4, 1988.
51. Theses of Mikhail Gorbachev's Speech at Plenary Meeting of the CPSU Central Committee, February 19, 1988 (Soviet Embassy, Washington, D.C., press release); M. Gorbachev, *Perestroika: New Thinking for Our Country and the World* (Harper & Row, 1987), pp. 19–21.
52. *New York Times*, September 26, 1988.
53. A. Toynbee, *Mankind and Mother Earth* (Oxford University Press, 1976), pp. 587–88.
54. *Los Angeles Times*, April 23, 1983.
55. L. R. Brown, *State of the World 1987: A Worldwatch Institute Report on Progress Toward a Sustainable Society* (Norton, 1987), pp. 212–13.
56. Cited by J. M. Kramer, "Environmental Problems," in J. Cracraft, ed., *The Soviet Union Today* (Bulletin of the Atomic Scientists, 1983), p. 153.

57. Ibid., p. 157.
58. Cited in H. F. French, "Industrial Wasteland," *World Watch,* November–December 1988, p. 22.
59. Kramer, op. cit., p. 160.
60. Cited in A. Pacey, *The Culture of Technology* (MIT Press, 1983), pp. 114–15, 178–79.
61. *Los Angeles Times,* September 13, 1984.
62. J. Strachey, *The End of Empire* (Praeger, 1964), p. 55.
63. J. Gay, "The Patriotic Prostitute," *The Progressive,* February 1985, p. 34; and *New York Times,* March 30, 1989.
64. *New York Times,* May 12, 1988.
65. *New York Times,* September 4, 1988.
66. *New York Times,* October 18, 1988.
67. Cited in M. Mies, *Patriarchy and Accumulation on a World Scale* (Zed, 1986), p. 189.
68. Ibid., p. 190.
69. M. Molyneux, "Mobilization Without Emancipation?" in R. R. Fagen, ed., *Transition and Development* (Monthly Review Press, 1986), pp. 280–302.
70. *Los Angeles Times,* March 14, 1987.
71. *The State of the World's Women, 1985: UN Decade for Women* (Nairobi, 1985), pp. 3, 17.
72. *New York Times,* July 17, 1985.
73. Aristotle, *The Politics,* 1253 b.
74. L. Harris, *Inside America* (Vintage Book, 1987), pp. 17–21, 122.
75. *Los Angeles Times,* June 16, 1988.
76. *New York Times,* March 26, 1986; July 2, 1988.
77. *Los Angeles Times,* January 22, 1987.
78. *New York Times,* July 1, 1987.
79. Brown, op. cit., p. 213.
80. Harris, op. cit., pp. 8–10.
81. *Wall Street Journal,* February 3, 1989; *Los Angeles Times,* October 9, 1988.
82. *Time,* May 25, 1987, p. 15; M. Stevens, *The Insiders: The Truth Behind the Scandal Rocking Wall Street* (Putnam, 1987).
83. D. LaBier, *Modern Madness: The Emotional Fallout of Success* (Addison-Wesley, 1986). See also S. Berglass, *The Success Syndrome* (Plenum, 1986), and statistics in Harris, *op. cit.,* pp. 8–10, 33–42, 51–56.
84. Bellah et al., *op. cit.,* pp. 284, 285, 295, 296.
85. K. S. Karol, *The Other Communism* (Hill & Wang, 1967), p. 248.
86. *Los Angeles Times,* April 1, 1986; *New York Times,* December 20, 1985.
87. *New York Times,* March 16, 1987.
88. Cited in O. Schell, *Discos and Democracy* (Pantheon, 1988), p. 294.
89. Reported by correspondent Henry Kamm in *New York Times,* June 26, 1987.
90. McNeill, op. cit., p. 140.
91. Ibid., pp. 277–80.
92. Toynbee, *Experiences* (Oxford University Press, 1969), p. 241.
93. *New York Times,* May 25, 1946.
94. *New York Times,* March 26, 1985.
95. P. R. Ehrlich et al., "The Long-Term Biological Consequences of Nuclear War," *Science* 222 (1983), pp. 1, 299.
96. Acts 4:32–35.
97. *Sane World,* Summer 1987, p. 18.

4. HUMAN PROSPECTS

1. N. McKendrick, J. Brewer, and J. H. Plumb, *The Birth of a Consumer Society* (Indiana University Press, 1982), p. 77.
2. *Dollars and Sense*, September 1987, p. 5; *New York Times*, June 9, 1989.
3. *Los Angeles Times*, April 30, 1987.
4. *Los Angeles Times*, November 13, 1987.
5. *Dollars and Sense*, December 1988, p. 4.
6. *In These Times*, April 1–7, 1987, p. 9; *New York Times*, August 28, 1988.
7. K. Jaspers, *The Origin and Goal of History* (Yale University Press, 1953), p. 1.
8. W. Baldwin, "Creative Destruction," *Forbes*, July 13, 1987, p. 49.
9. Cited by G. Bronson, "Technology: Songs the Sirens Sing," *Forbes*, July 13, 1987, p. 340.
10. Ibid.
11. Cited in L. Gubernick, "Faces Behind the Figures," *Forbes*, July 13, 1987, p. 450.
12. *Economic Justice for All: Pastoral Letter on Catholic Teaching and the U.S. Economy* (National Conference of Catholic Bishops, 1986), passim.
13. *New York Times*, October 22, 1987.
14. *New York Times*, March 23, 1989.
15. Press release from TIAA-CREF, January 18, 1988. Wharton is a specialist on economic development who served in various governmental and foundation groups, also headed at different times Michigan State University, the State University of New York system, and the Rockefeller Foundation, and currently is chairman of the $60-billion pension system for higher education, TIAA-CREF. The above remarks are from his keynote address at the 70th Annual Meeting of the American Council on Education in Washington, D.C. Emphasis in original.
16. Cited in *New Perspectives Quarterly*, Summer 1988, p. 33.
17. *New York Times*, June 15, 1988; *The Observer*, May 22, 1988.
18. *New York Times*, December 18, 1987 (Leonard Silk column).
19. Cited in E. P. Hoffmann, "Gorbachev's Trade Reforms," *Bulletin of the Atomic Scientists*, June 1988, pp. 22, 23.
20. Cited in N. Cohn, *The Pursuit of the Millennium* (Oxford University Press, 1970), p. 199.
21. R. H. Heilbroner, *The Future as History* (Harper, 1959), p. 99.
22. S. Schram, *Mao Tse-tung* (Penguin Books, 1966), p. 333.
23. *Los Angeles Times*, June 27, 1987.
24. *Los Angeles Times*, May 29, 1988.
25. *New York Times*, October 10, 1988.
26. Ibid.
27. *New York Times*, June 25, 1987.
28. Cited in P. Flaherty, "Perestroika Radicals: The Origins and Ideology of the Soviet New Left," *Monthly Review*, September 1988, pp. 22, 26.
29. A. Aganbegyan, "New Directions in Soviet Economics," *New Left Review*, no. 169 (May–June 1988), p. 95.
30. *Los Angeles Times*, December 25, 1987.
31. *New York Times*, October 14, 1988.
32. *Los Angeles Times*, December 18, 1988.

33. Interview of Jerry Rawlings by Diane Sawyer, CBS News, "60 Minutes," April 24, 1988.
34. Susan George, *A Fate Worse than Debt* (Grove Press, 1988); and Susan George article in *Los Angeles Times*, September 25, 1988.
35. *New York Times*, December 6, 1982.
36. *New York Times*, August 8, 1988.
37. R. Prebisch, "North-South Dialogue," *Third World Quarterly* II (January 1980), pp. 15–18.
38. R. Debray, "From Kalashnikovs to God and Computers," *New Perspectives Quarterly*, Fall 1988, p. 43.
39. *The Guardian*, June 1, 1988.
40. *Los Angeles Times*, August 11, 1986.
41. *New York Times*, November 2, 1988.
42. *The Nation*, January 2, 1989.
43. *New York Times*, January 9, 1980.
44. *Los Angeles Times*, October 15, 1987.
45. Cited by Abba Eban in *New York Times*, February 24, 1988.
46. *Los Angeles Times*, September 14, 1987.
47. Cited in J. A. Bill, "Resurgent Islam in the Persian Gulf," *Foreign Affairs*, Fall 1984, p. 111.
48. Cited in R. Wright, *Sacred Rage: The Wrath of Militant Islam* (Simon & Schuster, 1985), pp. 285–86.
49. R. Shaull, *Heralds of a New Reformation: The Poor of South and North America* (Orbis Books, 1984), p. 126.
50. M. Tangeman, "Liberation Theology Takes New Forms in Asia and Africa," *Latinamerica Press*, January 29, 1987, pp. 7–8.
51. Cited in R. Benne, *The Ethic of Democratic Capitalism* (Frontier Press, 1981), p. 4.
52. Shaull, op. cit., pp. 128–34. See also K. C. Ellis, ed., *The Vatican, Islam, and the Middle East* (Syracuse University Press, 1987).
53. Cited in N. Cohn, op. cit., p. 244.
54. Cited in P. N. Siegel, *The Meek and the Militant* (Zed Books, 1986), p. 6.
55. *New York Times*, December 25, 1986.
56. Mary E. Clark, "Meaningful Social Bonding as a Universal Human Need," in John Burton, ed., *Human Needs Theory and Conflict Provention* [*sic*] (St. Martin's Press, forthcoming). For a broader presentation of this view, see the indispensable multidisciplinary study by biologist Mary E. Clark, *Ariadne's Thread: The Search for New Modes of Thinking* (St. Martin's Press, 1989).
57. Heilbroner, *An Inquiry into the Human Prospect* (Norton, 1974), pp. 140, 141.
58. *New York Times*, February 4, 1980.
59. R. N. Bellah et al., *Habits of the Heart* (Harper & Row, 1985), p. 284.
60. Cited in R. Nelson, "When Civilization Ran Aground Aboard the Oil Tanker in Alaska," *Los Angeles Times*, April 6, 1989.
61. C. Bettelheim, *Class Struggles in the USSR: First Period 1917–1923* (Monthly Review Press, 1976), p. 331.
62. H. Kissinger, *The White House Years* (Little, Brown, 1979), p. 1063.
63. S. N. Kramer, *The Sumerians: Their History, Culture, and Character* (University of Chicago Press, 1963), p. 123.
64. Cited in G. Wint, *The British in Asia* (Institute of Pacific Relations, 1954), p. 18.

65. Press Release Soviet Embassy, Washington, D.C., February 19, 1988.
66. *New York Times*, November 14, 1988.
67. Cited in J. Schell, "A Better Today," *New Yorker*, February 3, 1986, pp. 47–67.
68. The description is by Professor Josef Silverstein of Rutgers University. *New York Times*, September 10, 1988.
69. Statement by Zwelakhe Sisulu, son of the imprisoned African National Congress leader, Walter Sisulu, cited in *Los Angeles Times*, April 7, 1986.
70. Cited in H. S. Commager, "The Revolution as a World Ideal," *Saturday Review*, December 13, 1975, p. 13.
71. Text of Nyerere's speech in *The Nation*, June 8–15, 1978.
72. J. M. Keynes, *Essays in Persuasion* (Harcourt, 1932), pp. 371–73.

5. WORLD HISTORY FOR THE TWENTY-FIRST CENTURY

1. *Bulletin of the Atomic Scientists* (March–April 1996), pp. 17–18.
2. *Los Angeles Times*, November 5, 1989; *New York Times*, December 8, 1992.
3. Cited by R. Lopez, *The Birth of Europe* (Evans, 1967), p. 25.
4. M. Nicolson, *Science and Imagination* (Cornell University Press, 1956), p. 15.
5. Ibid., p. 220.
6. Professor Lawrence Fuchs of Brandeis University, cited in *New York Times*, October 2, 1984.
7. Cited in L. Gubernick, "Faces behind the Figures," *Forbes*, July 13, 1987, p. 450.
8. *World Watch Institute Reports, State of the World, 1987, 1989* (Norton, 1987, 1989), pp. 213, 194.
9. Cited by J. Rifkin, *The End of Work* (Putnam, 1995), p. 221.
10. P.R. Ehrlich et al., "The Long-Term Biological Consequences of Nuclear War," *Science* 222 (1983), pp. 1, 299.
11. *New York Times*, September 21, 1995.
12. *Business Ethics*, March–April 1992, p. 30.
13. CBS television network, "The Twenty-First Century," May 21, 1967.
14. D. Suzuki, "Technology, Science, and Preparation for War," in T.L. Perry and J.G. Foulks, eds., *End the Arms Race* (West Vancouver, Canada: Soules Book Publishers, 1986), p. 327.
15. J. Cousteau, "Consumer Society Is the Enemy," *New Perspectives Quarterly* (Summer 1996), p. 48.
16. Ibid., p. 49.
17. *New York Times*, June 6, 1996.
18. *New Perspectives Quarterly* (Spring 1989), pp. 4–8.
19. Cited in A. Pacey, *The Culture of Technology* (MIT Press, 1983), p. 114.

INDEX

abortion, 169, 171
acid rain, 202
advertising, 136–37, 145
Afghanistan, 226, 249
Africa, 102, 104, 105, 107, 156,
   221
  colonization of, 120–21, 139
  women of, 167
African National Congress,
   247
Aganbegyan, Abel, 216–17
Agricultural Revolution, 185,
   186, 196
agriculture:
  in capitalist societies, 94,
   101–2
  genetic engineering and, 135
  in kinship societies, 28, 29
  in medieval Europe, 93
  slash-and-burn, 31–32, 33,
   47, 48, 78
  in Third World, 139–40
  trade and, 105
  in tributary societies, 47–48,
   51, 68
  variety of crops used in, 48
"agrobiology," 149
Alaskan oil spill, 238

alchemy, 112
alcohol, 28, 29, 30, 50, 106, 151
Alexander the Great, 241
Algeria, 219
Allende, Salvador, 231
alphabet, 90, 196
American Indians, *see* Native
   Americans
animals:
  kinship societies and, 31
  trading of, 104–5
  tributary societies and, 64
  use of tools by, 17–18
Aquino, Corazon, 69
Arabi, Ahmed, 119
Arabs, 58, 83, 101, 104
Arawak people, 96
Argentina, 69
Aristotle, 77, 97, 172, 174
Arkwright, Richard, 114, 115
Armenia, 66
arms manufacturers, 180–81
arms race, 184–85
Asia, 105, 109, 139
Asmat people, 39
Asoka, Emperor of India, 84
astronomy, 113
Athens, ancient, 70

atomic weapons, 115, 133, 147,
        183, 184–85
Australia, 122, 126
Australian aborigines, 20, 28,
        33, 35, 105–6
Australopithecus, 18, 20, 185
Avineri, Shlomo, 227
axial period, 198, 231–32
Aztecs, 57, 80, 96, 101, 106,
        118, 145, 241

Babylon, 84
Babylonia, 66
Bacon, Francis, 58, 99, 159, 187
Bahamas, 96
Baird, Father, 37
Balkans, 6
Balkans Since 1453
        (Stavrianos), 6
Ball, John, 208, 232
Bandura, Albert, 40
Banerjea, Surendranath, 123
Bantu farmers, 25, 28, 36
basic Christian communities
        (BCCs), 229, 231
Beijing (China), 194
Bellah, R. N., 138
Bemba people, 50, 54, 55,
        78–79
Ben Bella, Mohammed, 219
Berbers, 101
Berlin Conference, 120
Bevan, Aneurin, 126
Bill, James, 228
biology, 112, 113
birth control, 45, 49, 160, 162,
        169
birth rates, 29, 154
Bloomington, Minn., 193
Boesky, Ivan, 175

Bolshevik revolution (1917),
        129–30, 161, 209, 223
Bondarev, Yuri, 212
Boran people, 39
Boulton, Matthew, 87, 192
Brazil, 106, 142, 155, 229
Brezhnev, Leonid, 148–49, 150,
        211, 215
Briggs, Asa, 126
bronze, 83
Bronze Age, 52
Brown, Lester, 201, 202
Buddha, 198
Buddhism, 59–60, 226, 232
*Bulletin of the Atomic
        Scientists,* 248
burial, 74
Burke, Edmund, 126
Bureau of Antiques
        Administration, 74
Burlatsky, F. M., 212
business cycles, 137, 147
Butz, Earl, 155

Cámara, Helder, 231
Canada, 5, 107, 116, 146,
        184
capitalist societies, 10–11,
        87–187, 191, 239
    agriculture in, 94
    alternatives in, 198–208
    commercial stage of, 94,
        95–111, 191
    "creative destruction" of, 89,
        91, 133, 138, 175, 177,
        195–96
    cultural imperialism of, 145,
        194
    ecology and, 153–59, 202–3
    gender relations of, 160–72

global nature of, 87, 95, 104,
  111, 112, 116–17, 136–37,
  146, 191, 205
high-tech stage of, 95, 191
industrial stage of, 95,
  111–32, 191
legacy of, 236–37
profit motive in, 89, 94–95,
  96, 102, 175, 191, 238
rise of, 91–94
social relations of, 172–79
stages of, 95, 191
vs. tributary societies, 124
war in, 179–87
  see also Commercial
  Capitalism; High-Tech
  Capitalism; Industrial
  Capitalism
Caribbean, 165
Carnegie, Andrew, 87, 116,
  136, 137
cartels, 116–17, 136
Cartwright, Edmund, 114
castes, 61
castration, 69
Castro, Fidel, 225
Catherine the Great, 69, 76
Catholic Church, 170, 226,
  228–33
cattle industry, 156
cedars of Lebanon, 64–65
Census Bureau, U.S., 148, 203
Central America, 6
Challenger, USS, 153
Chang Hsieh, 96
Charles II, King of England, 97
Chartres Cathedral, 62
chemistry, 98, 112–13
Cheyenne Indians, 39
Chiang Kai-shek, 211

Chien Lung, Emperor of China,
  103
China, 57, 58, 59–61, 69, 74, 84,
  91, 92, 95–96, 109, 120, 123,
  159, 177–79, 193–94, 210–11
  class structure of, 72–73
  consumerism of, 177–78
  coolie labor of, 124–25
  Cultural Revolution of, 177
  emigration from, 104
  Europeans as viewed by,
    123, 194
  foreign investment in,
    193–94
  Great Britain and, 118–19
  inventions in, 58
  inwardness of, 59–61, 95–96,
    100, 103
  national identity in, 59–61
  private enterprise in, 60–61
  trade by, 104
  uprisings in, 75, 248
  women in, 70, 169
China Youth News, 178
Chin Shih Huang-ti, Emperor
  of China, 74
Churchill, Winston, 130
cities, 52, 57, 64, 140
clitoridectomy, 69
clothing, 192
coal, 157
cocoa, 105, 119
coffee, 105
coinage, 90, 196, 197
colonialism, 102–4, 113
  in Africa, 120–21
  conflicts of, 128
  raw materials and, 102–3
  rise of, 102–4, 118, 120–21
Colt, Henry, 114

Columbus, Christopher, 58,
    95–96, 106, 134
combines, 116
Commercial Capitalism, 94,
    95–111, 135, 191
  creativity in, 95–105
  definition of, 94, 96
  destruction by, 105–11
"Commission on the Year
    2000," 174
Communist Party, 210
"company countries," 143
"company towns," 143
computers, 134, 136, 149
Confucius, 61, 197, 198
Congress, U.S., 184
Connecticut Yankee in King
    Arthur's Court (Twain), 62
conquistadors, 58, 96, 101, 106,
    118, 185, 241
Constitution, U.S., 138
consumerism, 191–95, 237
coolie labor, 124–25
Copernican system, 113
copying machines, 148
Cortés, Hernando, 96
cottage industries, 94, 102, 109
cotton industry, 114, 115–16,
    122, 160–61
crafts, 51–52
creation myths, 13
Crick, Francis H. C., 134
Crompton, Samuel, 114, 115
Cromwell, Oliver, 240
Crow Indians, 30, 36, 39
Cuba, 225
culture:
  high, 55–56, 63
  low, 55–56
Czechoslovakia, 159

Darby, Abraham, 114
"Dark Ages," 91
Darwin, Charles, 113
Decade for Women, United
    Nations, 171
Declaration of Independence,
    247
de Gaulle, Charles, 195
Demosthenes, 70
Deng Xiaoping, 61, 177–78, 210
depression, psychological, 176
deserts, 66–67
Diamond, Stanley, 27, 43, 249
Díaz, Bernal, 57–58, 80–81
diseases:
  in Africa, 107
  control of, 117–18
  genetic engineering and, 135
  germ theory of, 113
  to New World, 58–59, 105–6
  tropical, 110
divorce, 162, 164
Domesday Book, 93
"Domesday Clock," 248–49
Dostoyevsky, F. M., 77
Downey, Thomas J., 201
Drucker, Peter F., 141, 147

East Indies, 109
ecology:
  capitalist societies and,
    153–59, 202–3
  kinship societies and, 30–33
  tributary societies and,
    64–68
Economist, 138
education:
  in kinship societies, 28, 36
  in tributary societies, 56, 77
  women's studies, 164

Egypt, 119
  ancient, 57, 64, 77, 79, 116
Einstein, Albert, 184, 186, 189
Eisenhower, Dwight D., 115,
  181
electoral politics, 242–45
Emerson, Ralph Waldo, 14
energy, consumption of, 32,
  158–59
Equal Rights Amendment, 164
Eskimos, 20, 23, 28, 29, 31, 33
Essay on Civil Government
  (Locke), 247, 248
Ethiopia, 66, 120
Eurasian civilizations, 58–59
Europe:
  China's view of, 99, 123
  economic union of, 207
  economies of, 147
  emigration from, 118, 122
  empires of, 128, 138, 180
Europe, Eastern, 209
Europe, Western, 119
  exploration by, 95–96
  navies of, 128
  rise of capitalism in, 91–95
  uprisings in, 110–11
evolution, 113
exploration, 95–96, 100
extermination camps, 183
Exxon, 238

famine, 140
farms, 148, 200–201
feminism, 164–65, 171
Fenton people, 39
fertilizers, 113, 135, 139
feudalism, 57, 92, 100, 179
fire, 18, 19
firearms, 107, 108, 133, 180

First World, 135, 142, 146–48,
  152
Flanders, 111
floods, 53
Florence (Italy), 111
food, consumption of, 173
food-gathering societies, *see*
  kinship societies
food production, 140, 154
Forbes, 199, 207
forests, 64–65, 155–56, 157
France, 62, 113, 117, 158,
  161–62, 180, 195, 219
  and NATO, 195
  and North Africa, 101, 119,
    120, 144, 219
Francis I, King of France,
  100–101
Francis Ferdinand, Archduke
  of Austria, 128
French Revolution, 62, 113,
  161–62
Freuchen, Peter, 20, 29, 31
Fuentes, Carlos, 226

Galileo Galilei, 98
Gama, Vasco da, 95, 96
Gandhi, Indira, 69
Gandhi, M. K., 219, 238
gender relations:
  in capitalist societies,
    160–72
  in kinship societies, 25, 26,
    33–34, 40, 68
  in tributary societies, 68–71
*General Theory of
  Employment, Interest, and
  Money* (Keynes), 13
genetic engineering, 134–35,
  136

genocide, 183, 185, 238
"geocide," 238
Germany, 83, 113, 119
Germany, Federal Republic of
    (West), 158
germ theory, 113
Ghana, 218–19
glasnost, 212, 216, 244
"global assembly line," 167–68
"global office," 168
gold, 96, 102, 183
Goldie, Sir Charles, 120
Gorbachev, Mikhail, 138, 179,
    208, 244, 245
  economic policies of, 149,
    150
  perestroika and, 151–52,
    195, 199, 211–18
grain, 48, 104
Great Britain, 74–75, 112, 117,
    180, 184, 194, 206, 220
  China and, 118–19
  Egypt and, 119
  empire of, 5, 101, 106, 120
  India and, 101, 120, 121–24,
    144, 219, 242
  industries of, 111
  revolution in, 110, 111, 126
  Spain and, 103
Great Depression, 3–5, 127,
    131, 147, 153, 204, 207, 209,
    211
Greece, 57, 90, 93, 241
"greenhouse effect," 157
"green revolution," 135, 139
Grey, Sir Edward, 128
gross national product (GNP),
    135, 138, 149, 202, 223
guilds, 94, 102, 109
gunpowder, 58

Habits of the Heart (Bellah, et
    al.), 138, 176, 238
Hadza people, 33
Haiti, 155, 226
Hammurabi, King of
    Babylonia, 66
Hammurabi's Code, 69
Hangchow (China), 80
harems, 69
Hargreaves, James, 114
headhunters, 39
height, 74–75
Heilbroner, Robert, 235
Heisenberg, Werner, 9
Herodotus, 241
Hero of Alexandria, 90
Hesiod, 61
High-Tech Capitalism, 95,
    132–53, 191
  creativity in, 133–38
  destruction by, 139–52
  military and, 133
  problems of, 133, 137–38,
    152–53
Hinduism, 77, 101
Hiroshima (Japan), 84, 115,
    133, 183
historians, role of, 5, 13–14
history:
  class bias of, 62–63, 75
  cycles of, 90, 91
  global approach to, 7–8, 13
  need for new approach to,
    8, 9, 13
  prehistory and, 9–10
  uses of, 5, 11–14
  Western bias of, 6–7, 13,
    57–59
Hitler, Adolf, 183, 185
Ho Chi Minh, 219

Homo economicus, 159, 238
Homo erectus, 18
Homo sapiens, 18, 19, 22, 38,
    185, 187, 192, 235, 238, 249
Hoover, Herbert, 211
Hopi Indians, 39
horses, 83, 93
Houphouët-Boigny, Félix, 219
House Ways and Means
    Committee, 201
Hsüan Tsang, 59
human nature, 9–10, 174, 191,
    235
    adaptability of, 18, 40–41,
      186
    aggression and, 38, 40
    destructive aspect of, 31
    festivity and, 80
    materialism and, 21
    revolutionary aspect of, 17
    selfishness vs. sharing in,
      36, 39, 235
Hungary, 159
Huns, 83, 84
hunting-gathering societies,
    see kinship societies
Hutterites, 49

Ice Ages, 18–19, 46, 186
Incas, 57, 101, 106, 241
incest taboos, 22
indentured servants, 103, 110
India, 57, 58, 69, 84, 109,
    155
    British rule in, 101, 120,
      121–24, 144, 219, 242
    coolie labor of, 124–25
    textile industry of, 115, 122
    women of, 165–66
Indonesia, 146

Industrial Capitalism, 95,
    111–322, 136, 191
    contradictions of, 121
    creativity in, 112–21
    crises of, 127–32
    definition of, 112
    destruction by, 121–27
    "long waves" of, 114
    industrial production levels:
      in France, 119
      in Germany, 119, 128
      in Great Britain, 119, 128
      in Japan, 147
      multinationals and, 137
      in Soviet Union, 131, 148
      in U.S., 119, 147
      of world, 117, 119, 131, 135,
        191
Industrial Revolution, 99, 112,
    117, 119, 121, 160
    political views of, 126–27
    working conditions of,
      125–26, 172
Industrial Workers of the
    World, 4, 5
infanticide, 29, 45, 70
infant mortality, 148
*Inquiry into the Human
    Prospect* (Heilbroner), 235
Institute for International
    Economics, 206
insulin, 135
insurance industry, 168
International Labor
    Organization, 166
International Monetary Fund,
    218, 221
intifada, 227
Iraq, 67
Ireland, Republic of, 147, 168

iron, 58, 83, 90, 114, 196
irrigation, 51, 65–66, 68, 78, 135, 139
Islam, 226–28, 230
Israel, 226–28
Italy, 104, 119, 147, 179, 180
Ivan the Terrible, 86

Japan, 130, 132, 145, 211
  economy of, 147, 149, 209
Java, 123
Jefferson, Thomas, 247
Jenkins, Robert, 103
Jews, 183, 185
John Paul II, Pope, 230, 232
joint-stock companies, 102, 109, 111, 112, 116, 135

Kalahari Desert, 24
karma, 77
Kay, John, 114
Keltie, Sir John, 119
Kennedy, Paul, 8, 10
Kentucky Fried Chicken, 194
Keynes, John Maynard, 13, 90, 199, 204–5, 207, 250
Khrushchev, Nikita, 148, 149, 211
Kim Il Sung, 225
kinship societies, 10, 15–41, 45, 191
  agriculture and, 28, 29
  communal nature of, 19–20, 21–23, 26, 35–37, 38, 73–74
  definition of, 19–20
  disintegration of, 28, 30, 36, 81
  ecology and, 30–33
  food gathering by, 20–21, 22, 23, 25, 26–27, 29, 33, 34

gender relations of, 25, 26, 33–34, 40, 68
hunting in, 20–21, 22, 25, 26–27, 31, 33, 34, 38
legacy of, 235
predominance of, 20
reasons for, 21–22
resource depletion and, 29
similarity of, 23–24
social relations of, 35–38
stability of, 27–28, 29, 64
war in, 38–41, 81–82
work in, 26, 37–38, 49
Kipling, Rudyard, 114, 120
Kitchener, Horatio, 86
Koch, Ed, 174
Koch, Robert, 117
Korea, Democratic People's Republic of (North), 225
Korea, Republic of (South), 142, 166
Korean War, 6, 135, 142
Kuikuru Indians, 49
Kung Bushmen, 23–28, 36, 39, 40, 49, 54, 55, 81
  communalism of, 26–27
  diet of, 25
  gender roles of, 25, 26
  health of, 25–26
  labor by, 26, 37–38
  social life of, 27

LaBier, Douglas, 176
land enclosures, 94, 101–2, 109, 139
language, 19
Latin America, 106, 120–21, 167, 221, 229–31
Latvia, 213
Laue, Theodore Von, 139

Lavoisier, Antoine, 112–13
Leacock, Eleanor, 34
Lebanon, 64–65, 226
Lee, Richard B., 37
Lei Feng, 177–78
leisure, 38, 172
le Jeune, Paul, 34, 35
Lenin, V. I., 129–30, 131,
    239–40
Lenin Agricultural Academy,
    149
Levellers, 110
Lewin, Moshe, 130
Liberia, 120
Libya, 119
Liebig, Justus von, 113
Lifton, Robert Jay, 187
Li Laikam, 194
Lippmann, Walter, 115
Lishen Toy Factory, 194
Lister, Joseph, 117, 118
Locke, John, 247, 248
Lockheed, 184
Long, Huey, 212
López Portillo, José, 145
Lowdermilk, Walter C., 67
Luce, Henry, 137
Luddites, 220
Luther, Martin, 232
Lysenko, T. D., 149

Macaulay, Thomas B., 144
machismo, 170
McNeill, William H., 115,
    179–80
Malaysia, 31–32, 40
malnutrition, 71, 140, 173, 222
Manhattan Project, 183
Mao Tse-tung, 60, 177, 193,
    194, 210, 240

mapmaking, 98
Marcos, Ferdinand, 146
Mareckova, Lena, 159
Martino, Renato R., 173
Marx, Karl, 124, 204, 209
Marxism, 60, 127, 132, 138,
    230–31
Massachusetts Institute of
    Technology, 184
materialism, 21, 35, 37, 114,
    185
Mayas, 57, 74
meat, in kinship society diets,
    33
Mencius, 61
Menshikov, Stanislav, 150
mercantilism, 102
Mercer, Robert E., 87
merchant class, 92, 93–94, 100,
    197
Mesopotamia, 55, 65, 69, 82,
    241
metallurgy, 51–52, 68, 90, 196
Mexico, 111, 116, 140
Michnik, Adam, 245
Micmac Indians, 37
Middle East, 49, 111, 226–28,
    230
military-industrial complex,
    115, 133, 181, 184
Mindanao Island (Philippines),
    38
Ming dynasty, 100
mining, 93, 98
Mitzna, Amram, 227
Moguls, 84, 118
monarchies, 100
Mongols, 83–84, 86, 242
monks, 92–93
Monopoly, 194–95

Moore, George, 199
Moslems, 83, 99, 100, 101,
    226–28
Mulroney, Brian, 184
multinational corporations,
    135–37, 139, 142–44, 146,
    167, 168, 200, 201, 202
Müntzer, Thomas, 232

Nabokov, Peter, 32
Nagasaki (Japan), 84, 133, 183
Napoleon I, Emperor of
    France, 162
Naskapi Indians, 34, 70–71
National Academy of
    Engineering, 199
National Conference of
    Catholic Bishops, 200
Native Americans:
    destruction of, 28, 58, 105–7,
        118
    differences among, 39
    ecological attitude of, 31, 32
    social relations of, 35, 73–74
natural resources, 202
natural selection, 113
navigation, 98
Nazis, efficiency of, 183–84
Nehru, Jawaharlal, 132, 144,
    219
Neolithic Age, 90
New Deal, 211, 216
New France, 107
New Guinea, 39
Newton, Sir Isaac, 113
New York, N.Y., 173–74
New York Daily Tribune, 124
Nicaragua, 169–70
Nile river, 51, 52, 66
Nixon, Richard M., 240

nomadism, 20–21, 29, 35, 46,
    48, 83–86
    war and, 83–86, 92, 183, 185
North Africa, 101, 119, 120, 144
North Atlantic Treaty
    Organization, 195
Novum Organum (Bacon), 99
"nuclear winter," 152, 185
Nye, Joseph, 204
Nyerere, Julius, 247

obnovlenie, 214
oil, 135, 150, 151, 157, 238
Old Horn, Dale, 36
"omnicide," 185, 238
Opium War, 118–19
Organization for Economic
    Cooperation and
    Development, 221
Organization of Petroleum
    Exporting Countries
    (OPEC), 135
Ottoman Empire, 97, 109, 111,
    118, 120
ozone, 156–57

Paiute Indians, 47
Pakistan, 146
Paleolithic era, 20, 22, 39–40,
    54, 172, 174, 192, 208, 235
Palestine Liberation
    Organization, 226–27
Pamyat, 212
Paris, Matthew, 85
Parker Brothers, 194–95
Pasteur, Louis, 113, 117
Pearce, Diane, 165
peasants, 55–56, 80, 90–91, 93,
    102
    colonialism and, 124

displacement of, 109–11, 125, 139–40
revolts by, 91, 110–11
perestroika, 151–52, 195, 199, 211–18
Perestroika (Gorbachev), 138
Perkin, W. H., 113
Perón, Evita, 69
Persian Empire, 120
Persian Gulf, 66
Peru, 111
pesticides, 135
Philippines, 69, 104, 146, 166–67, 226
Phillips, Kevin, 201
photo-copiers, 149
physics, 113
Plato, 52, 65, 79, 197, 208
plows, 68, 78, 90, 93, 196
Poland, 159, 245
Politics (Aristotle), 77
pollution, 32–33, 155, 158–59, 202
Polo, Marco, 80
population:
of Canada, 146
of China, 105, 125
control of, 29, 45, 154
of English colonies, 104
of English-speaking world, 104
of Europe, 117, 118, 153–54
of First World, 154
of India, 122, 125, 146
of Mexico City, 140
of Middle East, 49
of Spanish America, 106
of Third World, 154
of Western Europe, 93

of world, 30, 46, 52, 105, 113, 135, 153, 233
Portugal, 100–101, 106, 120
postindustrial society, 137
pottery, 51, 52
poverty, 71, 72, 148, 173, 200, 203
nature of, 80–81
Prebisch, Raúl, 223
prescription drugs, 158
priests, 53, 55
primate societies, 17, 22, 40
printing, 58, 99
private property, 69
prostitution, 166
Protagoras, 67
protectionism, 142
protein, 25
pseudoscience, 149
psychiatric hospitals, 175
public health, 117
Pugachev, Emelian, 75–76
Pyramids, 192, 192

race:
differentiation of, 19
in New World, 106
racism, 113, 114, 145, 166
radio, 145
railroads, 116, 119, 121, 180
Rakowski, Mieczyslaw, 225
Rawlings, Jerry, 218–19
Reader's Digest, 145
Reagan, Ronald, 175
Reddy, A. K. N., 144
religion, 53, 56–57, 101, 196, 198, 226–33
Republic (Plato), 79, 208
revolutions, 5, 177, 223–24
in England, 110, 161

in Europe, 6, 110–11
French, 62, 113, 161–62
in kinship societies, 27, 64
Russian, 129–30, 161, 209, 223
in Third World, 246
women and, 161–62, 169
Rhodes, Cecil, 120–21, 137
Rivera, Feliciana, 170
Roman Empire, 75, 84, 91, 93, 185, 232, 236
Roosevelt, Franklin D., 211, 212
Rousseau, Jean-Jacques, 63
Royal Society, 97
Russell, Bertrand, 234
Russia, tsarist, 75–76, 86
see also Soviet Union
Russian Revolution (1917), 129–30, 161, 209, 223

Sahlins, Marshall, 22–23, 38, 81
sailing, in ancient Egypt, 52
Sakharov, Andrei, 148, 151
Sandinista Defense Committee, 170
Sandinista Front, 169–70
Sargon I, King of Akkad, 64–65, 82, 84
Schumpeter, Joseph, 89, 95, 111, 175, 196, 199
science, 112, 115, 134, 149, 180
Scientific Revolution, 97, 99, 100, 112, 159, 186
Second World, 138, 148–52, 177
capitalism and, 193–95, 196
ecology and, 158–59
economic problems of, 148, 152, 209

Semai people, 40
Semang people, 31–32, 33
Sennacherib, King of Assyria, 84
Service, Elman, 20, 22, 40
service industries, 147
sexism, 166
"sex-tourism," 166
shamans, 53
Share-Our-Wealth Society, 212
Shaull, Richard, 231
Sherer, F. M., 199
shipbuilding, 98
Shipler, David K., 244
shipping, 116, 136
Shmelev, Nikolai, 214, 217
shopping centers, 193
Shultz, George P., 222
Siberia, 111, 116, 130
silver, 102, 104, 111
Six-Day War, 226
"60 Minutes," 218
slavery, 92, 103, 104, 110, 172
abolition of, 125
in Africa, 107–8
in antiquity, 57, 90
disease and, 58, 106
mortality of, 108, 110
in Roman Empire, 75
women and, 165
Smith, Adam, 107, 109, 175
Smuts, Jan Christiaan, 9, 15
Snow, C. P., 132–33
social Darwinism, 113–14, 120, 127, 201
socialism, 138, 151, 152, 208–18, 239–40
social reform, 117
social services, 165

societies:
  categories of, 10–11
  social organization
    important in, 17
soil erosion, 156, 202
Solidarity, 245
soup kitchens, 173
South Africa, Republic of, 28,
  246–47
Southeast Asia, 6, 104, 120,
  156, 166–67
Soviet Union, 127, 130, 132,
  133, 138, 159, 179, 195,
  244–45
  economy of, 148–52, 201,
    204, 209
  farms of, 217
  Five-Year Plans of, 131, 209
  military strength of, 150
  minorities of, 212–14
  New Left of, 214–15
  popular opinion in, 215–16,
    217
  republics of, 213
  technocrats of, 214
  working women of, 163,
    169
  see also Russia, tsarist
Spain, 100–101, 103, 106, 147,
  150
Spartacus, 75
speech, 18, 19
spice islands, 109
"stagflation," 146
Stalin, Joseph, 131, 149, 211
standard of living:
  of males vs. females, 164
  in Western Europe, 119
starvation, 71, 222
State of the World, 175

states, 69, 241
  European, rise of, 100
  functions of, 52, 53
  power of, 77
status:
  in Eskimo society, 23
  in kinship societies, 35, 39
  symbols of, 175
  of women, see gender
    relations
Stavrianos, L. S., youth and
  career of, 3–6, 7
steam engines, 90, 114, 116
steel industry, 136
Stefansson, Vilhjalmur, 23
stock market, 207
Strachey, John, 165–66
Strategic Defense Initiative
  (SDI; "Star Wars"), 184
stress, 175, 177
Suez Canal, 121, 124
suicide, 29
superstition, 97
Sweden, 158

Taiwan, 142
Tanzania, 33
Tasaday people, 38, 39
Tawney, R. H., 1
tea, 105, 119
technology, 17–18, 233–34
  in capitalist societies, 94–95
  "command," 115, 180, 183,
    184
  as cultural weapon, 144–45
  early, 17–18, 30, 38, 90
  gender relations and, 160
  high, 11, 33
  in Industrial Capitalism, 114
  and leisure, 38

in medieval Europe, 93
military, 83, 84, 86, 90, 115,
  118, 133–34, 182, 184
and social relations, 172
in tributary societies, 68,
  71–72, 78, 89–91
television, 145–46
termites, 17–18, 25
terrorism, 245–46
Thailand, 166
Thatcher, Margaret, 206, 225
Third World, 47, 89, 123, 133,
  139–46, 177, 196, 218–33
  cheap labor of, 136, 142, 146
  debt of, 141, 146, 221–22
  despotism in, 131, 222
  economic problems of, 138,
    139–44, 146, 203
  Europeanization of, 144
  First World and, 224
  industrialization of, 141–42,
    220
  nationalism in, 129, 139, 219
  Soviet Union and, 224
  women of, 165–72
Thirty Years' War, 100
Thomas Aquinas, Saint, 97
Tibet, 226
Tigris-Euphrates rivers, 51, 66
Timbuktu (Mali), 58
tobacco, 28, 29, 105
Tocqueville, Alexis de, 121
towns, beginnings of, 52
Toynbee, Arnold J., 8, 10, 132,
  152–53, 183
trade, 205
  fur, 107, 111
  Great Depression and, 131
  growth of, 94, 102–4, 110,
    111, 117

multinationals and, 137
silk, 104
slave, 107–8, 111–12
Trevelyan, Sir Charles, 122
tributary societies, 10, 20,
  43–86, 191
  agriculture and, 47–48, 51, 68
  bankruptcy of, 91
  capitalist societies vs., 124
  common features of, 56–57,
    58
  contradictions of, 61–63, 71
  creativity in, 62, 63
  definition of, 45, 52
  distribution of wealth in,
    71–72
  early examples of, 55
  ecology and, 64–68
  gender relations of, 68–71
  legacy of, 236
  productivity of, 48, 51–52,
    53, 55, 72
  sacred texts of, 56
  social relations of, 71–81
  stagnation of, 89–91
  structure of, 45, 52, 54, 56, 71
  styles of, 57
  transition to, 45–50, 51–54
  tributes in, 45, 52–53, 54, 109
  uprisings in, 75–77
  war in, 81–86
  work and, 50, 54, 56, 78–79
Tunis, 119
Turgenev, N. I., 76
Twain, Mark, 62
Tyler, Wat, 111

unemployment, 4, 30, 131, 200
  beginning of, 109–10
  in First World, 147, 222

in Ireland, 168
in Soviet Union, 217
unions, 161
United African Company, 120
United Nations, 138, 171
United States, 106, 116, 133,
    138, 147, 150, 159, 184,
    228
  Canada compared with, 5
  economic aid from, 142
  economy of, 147, 203
  family income in, 201
  Great Depression and, 131
  illegal aliens in, 140, 168
  Latin America and, 120–21,
    222
  Mexico and, 140
  New Left of, 215
  pollution in, 155
  rise of, 129
  rural poverty of, 148, 173
  working women of, 163
usury, 94

V-2 missiles, 134, 183
Vancouver (Canada), 3, 4
venereal disease, 28
Victoria, Queen of England,
    180
Vietnam, Socialist Republic of,
    169, 219, 225
Vietnam War, 135, 142, 166,
    219
Vikings, 106
villages, 21, 33, 124
  beginnings of, 48
  communal nature of, 50–51,
    54, 56
Virchow, Rudolf, 117
vodka, 151

wages:
  in Commercial Capitalism,
    94–95
  decline of, 147
  in First vs. Third World,
    143, 205
  in Industrial Capitalism, 117,
    125, 161
  in Ireland, 168
  minimum, 206
  in post-World War II boom,
    137
  of working women, 163–64,
    167–68
Walesa, Lech, 225
Wallerstein, Immanuel, 143–44,
    174
Walt Disney Company, 195
war:
  business of, 184, 186
  in capitalist societies,
    179–87
  in kinship societies, 38–41,
    81–82
  nomads and, 83–86, 92
  reasons for, 82–83
  technology of, 83, 84, 86, 90,
    179
  in tributary societies, 81–86
War of Jenkins' Ear, 103
"war socialism," 129
Watson, James D., 134
Watt, James, 90, 114, 192
Wedgwood, Josiah, 193
Weitz, Paul J., 153
West Indies, 103
Wharton, Clifton R., Jr., 203
wheat, 116, 150, 155
wheels, 52, 58, 90
White, Leslie A., 15

White, Lynn, Jr., 63
White, Robert, 199
Whitney, Eli, 116
Wilson, Woodrow, 129
Wisegroup Investment Ltd.,
    194
Wobblies, *see* Industrial
    Workers of the World
women:
  in Africa, 167
  cultural differences among,
    171–72
  in developed countries,
    160–65
  education and, 162
  enfranchisement of, 162, 242
  sexuality of, 162
  slavery and, 165
  in Third World, 165–72
  in work force, 163, 166–67
  see also gender relations
work:
  in capitalist societies, 172,
    200
  in kinship societies, 26,
    37–38, 49, 172
  in medieval Europe, 92

  in tributary societies, 50, 54,
    56, 78–79
World Conference on Women,
    171–72
World Food Conference, 173
World War I, 127–29, 130, 133,
    182
World War II, 115, 129, 132,
    133, 153, 183, 204, 216
World War III, 153
Worldwatch Institute, 158, 175,
    201
writing, 53, 55, 57, 62
Wu Chao, Empress of China,
    69

Yakovlev, A. N., 217
Yanomamo people, 40
yin and yang, 70
Yugoslavia, 159, 179

Zaslavskaia, Tatiana, 216
Zoroaster, 198
Zouaves, 101
Zulu proverb, 75
Zuni Indians, 39

L. S. Stavrianos was adjunct professor of history at the University of California, San Diego, and professor emeritus, Northwestern University, until his death in 2004. He earned a B.A. at the University of British Columbia and a Ph.D. at Clark University. He also taught at Queens University in Canada and at Smith College. Professor Stavrianos wrote more than a dozen books in Balkan and global history and published over four dozen articles. He received many awards including Guggenheim, Ford Faculty, Rockefeller Foundation, and Royal Society of Canada Fellowships.